Phenomenology in Adaptation Planning

Hendricus Andy Simarmata

Phenomenology in Adaptation Planning

An Empirical Study of Flood-affected People in Kampung Muara Baru Jakarta

Hendricus Andy Simarmata
University of Indonesia
Jakarta, Indonesia

ISBN 978-981-10-5495-2   ISBN 978-981-10-5496-9 (eBook)
DOI 10.1007/978-981-10-5496-9

Library of Congress Control Number: 2017947043

© Springer Nature Singapore Pte Ltd. 2018
This work is subject to copyright. All rights are reserved by the Publisher, whether the whole or part of the material is concerned, specifically the rights of translation, reprinting, reuse of illustrations, recitation, broadcasting, reproduction on microfilms or in any other physical way, and transmission or information storage and retrieval, electronic adaptation, computer software, or by similar or dissimilar methodology now known or hereafter developed.
The use of general descriptive names, registered names, trademarks, service marks, etc. in this publication does not imply, even in the absence of a specific statement, that such names are exempt from the relevant protective laws and regulations and therefore free for general use.
The publisher, the authors and the editors are safe to assume that the advice and information in this book are believed to be true and accurate at the date of publication. Neither the publisher nor the authors or the editors give a warranty, express or implied, with respect to the material contained herein or for any errors or omissions that may have been made. The publisher remains neutral with regard to jurisdictional claims in published maps and institutional affiliations.

Printed on acid-free paper

This Springer imprint is published by Springer Nature
The registered company is Springer Nature Singapore Pte Ltd.
The registered company address is: 152 Beach Road, #21-01/04 Gateway East, Singapore 189721, Singapore

To my parent and brothers, wife, daughter, and son

Foreword

This is a book for professionals, scholars, and general audiences who are passionate about the phenomenology of flood risks and adaptation planning in Indonesia. This book is a rewritten dissertation that assesses local practices of living with and managing regularly and—due to climatically induced sea level rise—increasingly occurring flood events in Jakarta. The author, Mr. Simarmata, here pays particular attention to the way locals perceive and make sense of the regularly occurring floods and based on these sense-making processes design their coping and adaptation strategies that for many entail the decision to not only find ways of living with the floods but in addition making use of them for livelihood provision. It is this local "planning knowledge" as Simarmata calls it that stands at the center of his interest in reducing the gap between formal urban planning processes, Simarmata's area of professional expertise before returning to the university, and planning through everyday practice.

Mr. Simarmata identifies this discrepancy between city-level planning and local planning as the point of departure of his research by stating: "At the city level, the plan is to control the floods; at the community level, the plan is to live with them" (p. v). Both types of planning come with advantages and disadvantages. While the aim to control the floods understands the floods as potential serious threats that are likely to not only increase in occurrence but also in force, the aim to live with them understands their positive, life-bearing, economic and sociocultural potential. The former takes a long-term perspective, preparing for the potential worst; the latter allows a nearly playful, but possibly short-sighted, approach to seeing the beauty in floods for everyday life, inherently acknowledging that any form of adaptation planning is—as social constructivist thoughts on innovation development have taught us—always about simultaneously developing and implementing a plan and building a society. Simarmata assigns this discrepancy to different planning methods, different planning interests, and varying sources of knowledge.

For studying these in detail, he takes a phenomenological approach drawing on Alfred Schütz's concept of the lifeworld (Chap. 2), thus contributing to phenomenologically inspired planning theory (including discussions on self-reflective action in planning). Chapter 3 outlines assessment on the discrepancy between city-level government (formal) planning with a clear focus on climate proofing and the neighborhood (kampung)-level everyday experiential planning, followed by a discussion on the kampung (neighborhood) as location for local-level flood management (Chap. 4). In this chapter and based on his data, Simarmata then unfolds his assessment of the role of individuals' lifeworlds in local-level, everyday flood management. He explicitly argues that the lived experience accumulated in individuals' lifeworlds acts as a crucial source for practical planning knowledge, yet also clearly delineates the "zone of planning" and thus the spatial (territory and scale) as well as temporal (long-/short-term) reach of the adaptation planning. Given the poor of Kampung Muara Baru, one of the poorest and most affected areas in Jakarta, can actively make use of these lived experiences, Simarmata argues that they are not necessarily to be considered vulnerable. But instead, the poor have to be indeed understood as independent, self-organized, and self-reflected agents (Chap. 5). The intersubjectively shared meaning processes into which the local planning processes are embedded in and of which they are a result of have assured strong ownership in the local adaptation planning and implementation. That does not make it fit long-term climate change scenarios for the region, but it assures immediate local fit and applicability within the current context. The enforcement and institutionalization (Chap. 6) of the local, orally communicated, adaptation plan is assured through a range of local institutions, such as kerja bakti (joint voluntary work), arisan (social gatherings), and pengajian (regular Muslim prayers). Hendricus Simarmata here empirically assesses the socially constructed character—along the lines of Peter Berger and Thomas Luckmann—of everyday, local-level (informal) planning, from meaning construction to rule and role formation, as well as to the institutionalization and reification of the developed adaptation plan and practices. Local-level governance and change adaptation mechanisms are here assessed by documenting the intricate local patterns of self-organization—routinized, institutionalized forms of social organization with the aim to live with, rather than prevent or move away from, floods. Thus, Simarmata understands to shed a different light on floods and enable the reader to see floods through the eyes of the inhabitants of Kampung Muara Baru—as a hassle, yes, but also as an opportunity to achieve additional earnings. Yet, rather than romanticizing floods altogether, Simarmata also understands the limited character of this local-level adaptation planning. While it fits local realities and is literally owned by the people themselves, implementation rates are high, making it possible to live with the floods—as long as these do not exceed given degrees of force and height. Protection against higher/stronger floods nevertheless continues to remain with the authorities of Jakarta, exceeding the planning scope of the kampung.

The presented analysis is based on a unique wealth of empirical data, ranging from spatial data presented in maps to visual as well as substantial qualitative ethnographic and interview data collected by the author himself through the employment of a diverse range of different methods during altogether 12 months of empirical field research in Kampung Muara Baru. Conceptually, the assessment of local lifeworlds and their role in adaptation planning offers an innovative and so far underrepresented approach to change adaptation. It adds a local-level planning perspective, embedded into everyday life circumstances of the urban poor, to conventional approaches to flood management, largely focusing on technical climate change adaptation scenarios and models. Thus, and with the aim to assure that his phenomenologically inspired qualitative, ethnographic analysis is able to communicate with the largely more technocratic, managerial approaches to change adaptation, Simarmata, toward the end of his thesis, brings his findings together in what he himself describes as a locally embedded adaptation planning model. He states: "This model connects the lifeworld, reflective practices and institutionalization to an integrated planning cycle. This planning model not only addresses the problems that have been defined by the local people, but also increases their ownership of the planning product" (p. 165). While the use of the notion "model" can be debated here, Simarmata's work contributes an important and often neglected perspective to ongoing debates on change adaptation (urban) planning by shedding rich empirical light on the intricate patterns of self-organization as well as the discrepancies but also potential linkage points between local-level urban planning and city-level urban planning. The reflection on what this means in detail for Schütz's lifeworld concept, its empirical applicability and analytical usefulness, especially also in the Indonesian context, unfortunately is less developed in the thesis. Yet, these conceptual weaknesses, just as the challenges in the use of the English language, are outbalanced by the empirical wealth and analysis thereof. Furthermore, the degree of interdisciplinary integration (methodologically and conceptually) found in this book in quality substantially exceeds levels usually demanded of and found in PhD theses. It is here that Mr. Simarmata shows a high level of disciplinary understanding of rather applied/managerial urban planning processes for climate change adaptation *as well as* processes of meaning construction and patterns of social organization and governance enabling him to integrate the different disciplinary as well as epistemological approaches into an interdisciplinary perspective and analysis. I therefore recommend this book to be read by anyone who has interests in phenomenology of planning, adaptation to climate change, and disaster risk reduction, especially in the Southeast Asian region.

Professor of Sociology of Knowledge Anna-Katharina Hornidge
Head of Department of Social Science
Leibniz Centre for Tropical Marine Research (ZMT)
Bremen, Germany
25 April 2017

Preface

Phenomenology in planning can be understood as an aspect of philosophy studying into the experience of arranging a plan. It concerned with people's understanding of planning practices. Adaptation to environmental change is the longest practice that has ever been performed by humankind. It always evolves and thus will transform planning knowledge from generation to generation. This book presents the narratives of how flood-affected people in Kampung Jakarta use and inter-share their experiential knowledge to arrange adaptation planning. Inspired by the flood story of Kampung Muara Baru (KMB) Jakarta and lack of empirical evidence of people-based approach, this book offers a new model of people-centered adaptation planning.

Floods increase the vulnerability of the residents of flood-prone areas and affect their everyday lives. Minimizing the impact of floods requires an adaptation plan, especially at the community level. Moreover, cities and communities need a synergized response plan. A study based strictly on technical science would not determine the essential meaning of adaptation planning as experienced by a vulnerable people. The planning knowledge of adaptation that is generated by the flood experiences of the urban poor—the most vulnerable group—must be disclosed. We need to know more about the "lived experiences" of people's adaptation to floods and the meanings that these people ascribe to their everyday lives. Therefore, only a people-centered approach can ascertain how the urban poor adapt to the floods.

Drawing on Schütz's concept of lifeworld and 12 months of fieldwork in Kampung Muara Baru (KMB) Jakarta, this book offers an in-depth investigation on how individuals use their lifeworld as a basis for practicing and institutionalizing their community's planning for adaptation. I begin with the context and locus of the adaptation planning studied in the present research, and I formulate two subquestions: What kind of planning institutions is constituted in Jakarta? How does KMB represent the interplay between poor residents and regular floods? I then focus on the adaptation practices of the KMB people, especially their perceptions of their own flood-related vulnerability. Second, I identify the meaning of adaptation planning, and I examine the institutionalization and reification of the adaptation plan. I apply the lifeworld analysis to examine the concepts of vulnerability, adaptation, and planning. In addition to examining the secondary data, I collected primary data in the forms of participant observations, in-depth and semi-structured interviews, group interviews, historical transects, and focus group discussions. It is expected to provide a better understanding of the appropriateness of the lifeworld concept in planning practices and to extend the horizon of phenomenology in planning.

This study found that more than one kind of adaptation planning addressed flooding in Jakarta. At the city level, the plan was to control the floods; at the community level, the plan was to live with them. This divergence was caused by discrete departure points, different planning methods, and varying sources of knowledge. It thus interfered with the institutionalization of planning because the divergent worlds of city and kampung were not connected. The results showed that, as an agglomeration of kampungs, Jakarta should understand the relationship between floods and the urban poor within the kampungs. Even though KMB had the greatest flood risk and is the poorest settlement in the Penjaringan subdistrict, the recurrent floods do not discourage migration into and the spread of housing in KMB. Instead, flood incidents have become major inundation events because of KMB's high-density settlement and poor drainage system.

Based on the in-depth analysis of the lifeworld of the people affected by flooding in KMB, I found that the poor who lived in the flood-prone area were not always vulnerable. Lived experience is an important factor that makes a significant difference between the vulnerable and the adapters. Lived experience is a source of the practical knowledge that is useful in planning. The structure of the lifeworld delineated the zone of planning operations and adaptation practices and thus resulted in short-term perspectives, neighborhood scaling, and problem-solving orientation, rather than long-term, citywide scaling and visionary planning. The KMB people do not plan to stop or to mitigate the floods but to adjust their houses and surroundings to reduce the consequences of flooding and develop an evacuation pathway based on their lived experiences. Thus, their adaptation planning is locally embedded.

Because the intersubjectively shared meaning process has produced locally embedded planning, the findings showed a strong ownership of adaptation planning. The KMB people embodied the plan through the self-interpretation and self-reflection of what their predecessors and others had shared. Subsequently, they discussed the substance of the planning in order to make deliberate decision about the course of a series of social events, such as kerja bakti (working together in voluntary services), arisan (regular social gatherings), and pengajian (routine Muslim praying), and casual events, such as warung (small shop) talks and alley chats. Therefore, they arrived at a verbal plan that was never written. They preferred listening to reading and talking to writing. Even though the plan consisted of loose-fitting regulations without clear penalties for violations, they teased and made fun of those who did not accomplish their tasks or broke the oral agreement.

Based on the lifeworld analysis of flood-related vulnerability, adaptation, and planning as the embodiment of a people-centered approach, I created a model of locally embedded adaptation planning. This six-step protocol consists of identifying the adapters, compiling precedents, revealing and assessing the locally situated form of knowledge, facilitating the sharing process, establishing the oral consensus, and dividing the tasks and responsibilities.

This model contributes to the ongoing debate on adaptation planning in the context of climate change adaptation and disaster risk reduction as an alternative, insightful approach to linking coping strategies to adaptive capacity. In terms of flood management, the model of locally embedded adaptation planning explains the relationship between the flood responses that are needed and those provided by the community. The model thus strengthens the response strategies in flood management to prepare a resilient community. Finally, in terms of planning practices, this model incorporates the humanistic values into adaptation planning through lifeworld analysis.

Driven by the phenomenological study, this book does not go deep to debate Husserl's and Schütz's approach in defining intentionality of meaning generation, but explores more about social practices in organizing adaptation planning process at the micro level. It contributes to the development of phenomenology in planning, especially empirical practices in adaptation planning as well as meaningful interpretations of the experiences of people who practice adaptation planning and social institutions that embody the adaptation planning process. Even though the focus is only at individual practices and the interrelation among individuals, which would not address the multilevel urban environmental problems, this study can demonstrate that the different realms of planning domain occur and the emergence of converting tools to translate the planning knowledge between locals and experts is needed to ensure the inclusiveness of urban planning takes place.

On the scientific level, this book is expected to enrich planning theory and the practice of accommodating the dimensions of adaption to climate change, to investigate the appropriateness of the *lifeworld* concept in planning practices, and to provide lessons learned about institutionalization in the theory of planning as a social phenomenon. It provides findings that link the *lifeworld* and planning theory. The book also attempts to extend the horizon of phenomenology in the field of planning and to contribute to previous phenomenological research and qualitative methods. The contribution of this book to the theoretical development of the institutionalization process of adaptation planning is identified as follows:

1. This book offers the *lifeworld* approach in conceptualizing vulnerability to climate change. Previous studies of vulnerability emphasized regional and sociospatial approaches without paying much attention to the identification of vulnerable people. It aims to provide information from a perspective that shifts from regional to personal approach.
2. This book also provides Indonesian perspectives on how a vulnerable community plans to adapt to environmental changes. The study can enrich the results of previous studies regarding community-based adaptation planning for climate-related floods in developing countries, such as UNDP (2013), ADB (2013), and the International Institute for Sustainable Development (IISD) (2012). The typical characteristic of the Indonesian *kampung* can provide a different view of how community-based adaptation planning is conducted in several developing

countries, because kampung residents mainly come from different islands and ethnicities, compared to Malaysia, Brunei, or Singapore.
3. This book brings the *lifeworld* perspective into planning theory and practice. Previous studies on self-reflective action in planning have focused on industrialized countries such as the USA. Moreover, the experiences of developing countries can be examined as a discourse that provides the underlying context in which the boundaries of planning are established.
4. This book adds the perspective of the sociology of knowledge to the debate of a place- and people-based approach to adaptation to environmental changes. It is expected to provide empirical evidence that people in vulnerable places know how to adapt to changes. It thus aims to strengthen the emergence of a people-centered approach in urban development studies.
5. This book focuses on developing a micro-sustainable development pathway that will provide suggestions about how localities in the community need to be considered in the policy-making for citywide development, especially urban poverty and climate-related disasters.

On the societal level, this book intends to provide new findings that will help meet the challenge of institutionalizing adaptation planning practices in Indonesia. It will benefit local governments through creating an enabling environment for communities that have taken initiatives to develop their own adaptation strategies. The book is also expected to promote a new approach to institutionalism in planning practices. This book is also expected to help urban and planning theorists, academics, and researchers to develop a better understanding of locally embedded adaptation planning and how it is institutionalized in informal urban settlements, such as the *kampung*s in Indonesia. To a lesser extent, the findings of this research could also be used as a conceptual resource for urban policy-makers and practitioners in Indonesia to strengthen micro-level policy and planning strategies that deal with environmental changes in the unique urban settlements in Indonesian cities.

This book is divided into six main parts. The first chapter presents an introduction why this study is relevant to the current and future challenges of environmental risks in the urban development discourse. It also explains why lifeworld is used as the approach to understand the phenomenology of adaptation planning that is practiced by people of Kampung Muara Baru (KMB) Jakarta.

The second chapter discusses the recent debate of phenomenology in planning, exploring the linkage between knowledge and action, particularly in the process of adapting to environmental change, in the area where locally embedded adaptation planning takes place. I rationalize the application of Schütz's *lifeworld* theory in three intersecting areas: (1) the *lifeworld* perspective in identifying vulnerable people, (2) the structure of the *lifeworld* to generate the meaning of locally embedded adaptation planning, and (3) the analysis of *lifeworld* to explain the institutionalization process of adaptation planning.

In the third chapter, I discuss the contextual features of adaptation planning in Jakarta. I explain the planning institutions that are constituted in Jakarta regarding flood management and how government regulations and the initiatives of other stakeholders work to take responsibility for the effects of floods. I found that although different planning approaches have been taken, which focus mainly on flood infrastructure development at the city level, they have not been coordinated effectively. Next, I argue that urban adaptation planning in Jakarta should focus on community empowerment and rely on the local knowledge that is embedded in each community. The different realms of planning would produce inefficient and ineffective adaptation and potentially create other new problems.

The fourth chapter presents the context where interplay between the floods and the poor takes place. This is an analytical description of the nexus of the urban poor and the floods that are perceived as a major problem in urban development both globally and locally. It is followed by discussing flood-related vulnerability based on the perception of KMB people and analyzing the perceptions of three main elements of vulnerability: exposure, sensitivity, and adaptive capacity. I examine how these elements are used to identify the vulnerable and the adapters. I end the discussion by arguing that they have generated their own criteria to define flood-related vulnerability and to identify the vulnerable and the adapter.

The fifth chapter elaborates what the meaning of locally embedded adaptation planning is and how it is institutionalized in *kampung*. The discussion focuses on the lived experiences of individuals and groups in planning their housing management, evacuation and shelter strategy, flood infrastructure provision, and income generation. Following that, this chapter explains how their planning activities become habitualized actions and how they reciprocally typify their planning. I argue that locally embedded adaptation planning is planning knowledge that is produced by socially reflected adaptive practices through the shared meaning of lived experiences. This chapter discusses the presence of informal planners who transmit planning knowledge through several nonformal events in the community. The unwritten communication has played a significant part in producing the planning knowledge. Lastly the sixth chapter concludes by considering the interrelation between the results of the empirical findings and the phenomenology in planning, including its contribution to climate change adaptation, flood management, urban policies, and planning practices.

Jakarta, Indonesia   Hendricus Andy Simarmata

Acknowledgments

This book was initially developed during my doctoral studies in Bonn. Most earnest acknowledgment goes to my supervisors, Prof. Dr. Anna-Katharina Hornidge and Prof. Dr. Christoph Antweiler, who patiently advised and guided me during my doctoral study. Prof. Hornidge has introduced me to the Schütz's lifeworld and a treatise of Berger and Luckmann and sharpened my discussions and arguments in this book. Prof. Antweiler has introduced me to the discourse of knowledge. I would also like to acknowledge both Prof. Stephan Conermann and Prof. Conrad Schetter for their valuable comments and advises when they examined the earlier version of this book.

Institutional supports from Zentrum für Entwicklungsforschung (ZEF), Leibniz Center for Tropical Marine Research (ZMT) Bremen, the Directorate General of Higher Education of Indonesia (DIKTI), Universitas Indonesia (UI), the Indonesian Association of Urban and Regional Planners (IAP), and the International Secretariat of Global Change System for Analysis, Research, and Training (START), when I took my doctoral study. My sincere thanks to ZEF directors and doctoral program coordinator, the ZEF-2011 batch, ZEF researchers, and all staffs for the support and friendship. I am obliged to thank DIKTI for awarding me a 3.5-year scholarship of doctoral program through *Beasiswa Unggulan*. My gratitude to the rector, the former and recent head of the graduate school, and the head of and faculty members of the urban studies postgraduate program of UI for allowing me to pursue my doctoral degree. My sincere thanks to the president of IAP and director of START for the field research funding. Also, to various organizations, such as ISSC, Paris; BIARI, Providence; and BWPI, Manchester, that awarded me travel grants and gave me a chance to share parts of this book into their programs. All comments that arose during those conferences have enriched the discussion part of this book.

I would like to say my deepest gratitude to the residents of Kampung Muara Baru (KMB) and all my interviewees who have shared their tacit experiences and treated me and my research assistants like a friend during the fieldwork. To my research assistants and my colleagues in IAP, PRPW UI, HAS Advisory Group, and Studio Poltangan and Bogor, for helping me in seeking data and supporting the bookmaking process. I am equally thankful to John and Jenna for proofing my English writing.

Last but not least, I owe a debt of gratitude to my dear wife Selvie Maylina, my daughter Louisa, and my son Sebastian for helping through my first international publication with their patience and love. There are no words that could explain how blessed and proud I am to have them in my life. Finally, in the name of the Father, the Son, and the Holy Spirit, I praise Jesus Christ, my Savior. I would not be able to finish this book without Him.

Contents

1	**Adaptation Planning for Floods**...	1
	1.1 Flooding as a Continuous Problem in Coastal Cities	2
	1.1.1 Floods and the Urban Poor	4
	1.1.2 *Kampung* as the Study's Focus	6
	1.2 People as Subjects of Adaptation to Climate Change	9
	1.3 People-Based Adaptation Planning	12
2	**Lifeworld as the Domain of Adaptation Planning**	17
	2.1 Lifeworld and Planning	17
	2.2 Adaptation Planning: Community- and People-Based Approaches	24
	2.3 The Nexus of Vulnerability and Adaptation: A People-Based Approach..	28
	2.4 The Framework of People-Centered Adaptation Planning................	35
3	**Planning Institutions of Adaptation to Flood in Jakarta**	39
	3.1 Understanding Floods in North Coastal Jakarta	39
	3.2 City-Level Adaptation Planning: Infrastructure-Driven Approach..	44
	3.3 Unrecognized Knowledge of Adaptation Planning at Community Level	58
	3.4 Different Types of Adaptation Planning...	61
4	**Flood Experiences: "The Vulnerable" and "The Adapter"**	65
	4.1 Kampung Muara Baru (KMB): The Interplay Between the Floods and the Poor	66
	4.2 Factors Affecting the Floods in Kampung Muara Baru.....................	70
	4.2.1 Types of KMB Dwellers..	70
	4.2.2 Land and Housing Status	73
	4.2.3 Basic Infrastructure	78
	4.2.4 Formal and Informal Economies	79
	4.3 Increasing Flood Risks and Growing Population	80

	4.4	The *Lifeworld* of the People of Kampung Muara Baru	82
		4.4.1 Shifting Meanings of Floods: From Opponent to "Friend"	90
		4.4.2 Self-Sensing: A Different Lens in Perceiving Floods	92
	4.5	KMB People's Perception of Flood-Related Vulnerability	94
		4.5.1 Perceived Flood Exposure	94
		4.5.2 Perceived Flood Sensitivity	97
		4.5.3 Flood-Related Adaptive Capacity	98
		4.5.4 Adaptive Capacity of Agus, the Head of RT 15	100
	4.6	The Continuum of "The Vulnerable" and "The Adapters"	101
5	**Locally Embedded Adaptation Planning**		**105**
	5.1	Reflection on Experiences and Learning	106
		5.1.1 Scenario Planning for Evacuation	107
		5.1.2 Planning for Living Space Arrangements	112
		5.1.3 Planning for Neighborhood Infrastructure Provision	117
	5.2	Reflective Practice as a Source of Planning Knowledge for Adaptation	120
	5.3	Adapter as an Informal Planner	124
	5.4	Institutionalization and Reification of LEAP	130
		5.4.1 House Management Plan	131
		5.4.2 Evacuation Plan	132
		5.4.3 Another Plan	134
	5.5	Connecting to City Level: A Language of Planning Knowledge Converter Required	136
6	**Planning and Adaptive Knowledge**		**145**
	6.1	The Social World as a Domain of LEAP	146
	6.2	LEAP is a New Insight for Planning Debate	148
	6.3	LEAP is a People-Centered Adaptation	151
	6.4	Contribution to Flood Management	153
	6.5	Implications for Urban Policies and Planning Practices	154

Appendices	**157**
Appendix 1: Methodology	157
Research Design	157
Field Work Procedure	158
Appendix 2: Participant Information	166
Introduction	166
Participation	167
Expected Benefits	167
Appendix 3: Semi-structured Interview Guidelines	168
Planners	168
NGOs	169
Government Officers	170
Head of RTs/RWs	171
Appendix 4: Transcript Form	173

Appendix 5: The Occurrence of Tidal Floods (Banjir Rob)
from 2007 to 2014.. 173
Appendix 6: Social and Public Facilities in Kelurahan Penjaringan 175
Appendix 7: Superimpose Analysis with ArcGIS 9.0 175
Analytical Tools Using ArcGIS 176
Data Processing.. 177

Glossary 185

References 189

Abbreviations

AADMER	ASEAN Agreement on Disaster Management and Emergency Response
AAL	Annual average losses
ACCCRN	Asian Cities Climate Change Resilience Network
ACF	Action Contre La Faim
ADB	Asian Development Bank
AICP	American Institute of Certified Planners
AKP	Adaptation Knowledge Platform
AMS	American Meteorological Society
APSC	ASEAN Political Security Community
Bappeda	Badan Perencanaan Pembangunan Daerah (Regional Development Planning Agency)
Bappenas	Badan Perencanaan Pembangunan Nasional (National Development Planning)
BBC	British Broadcasting Corporation
BLHD	Badan Lingkungan Hidup Daerah (Environmental Management Agency)
BMKG	Badan Meteorologi, Klimatologi dan Geofisika (Indonesian Agency for Meteorology, Climatology and Geophysics)
BNPB	Badan Nasional Penanggulangan Bencana (National Agency of Disaster Management)
BPBD	Badan Penanggulangan Bencana Daerah (Local Disaster Management Agency)
BPPT	Badan Penelitian dan Pengkajian Teknologi (Agency for the Assessment and Application of Technology)
BPS	Biro Pusat Statistik (Central Bureau of Statistics)
BWPI	Brooks World Poverty Institute
CaR	City at Risk forum
CARE	Cooperative for Assistance and Relief Everywhere
CBA	Community-based adaptation

CBO	Community-based organizations
CCA	Climate change adaptation
CCCD	Commission on Climate Change and Development
CDRM	Community-based disaster risk management
CIP	Canadian Institute of Planners
CoBRA	Community-Based Resilience Assessment
CRisTAL	Community-Based Risk Screening Tool for Adaptation and Livelihood
DFID	Department for International Development
DIBI	Data and Information of Indonesian Disasters
DJPR	Directorate General of Spatial Planning
DKI Jakarta	Daerah Khusus Ibukota Jakarta (Special Capital Region of Jakarta)
DNPI	Dewan Nasional Perubahan Iklim (National Council of Climate Change)
DRI	Disaster Risk Index
DRR	Disaster risk reduction
FGD	Focus group discussion
GDP	Gross domestic product
HCMC	Ho Chi Minh City
IAP	Indonesian Association of Planner
ICLEI	International Council for Local Environmental Initiatives
IHDP	International Human Dimension Program
IISD	International Institute for Sustainable Development
IPB	Institut Pertanian Bogor (Bogor Agriculture Institute)
IPCC	Intergovernmental Panel on Climate Change
IRSA	Indonesian Regional Science Association
ISSC	IEEE Symposium on Computers and Communications
ITB	Institut Teknologi Bandung (Bandung Technology of Institute)
IUCN	International Union for Conservation of Nature and Natural Resources Jabodetabekpunjur Jakarta, Bogor, Tangerang, Bekasi, Puncak and Cianjur
JCDS	Jakarta Coastal Defense Strategy
JEDI	Jakarta Emergency Dredging Initiatives
JICA	Japan International Cooperation Agency
JICA-RI	Japan International Cooperation Agency Research Institute
JUFMP	Jakarta Urgent Flood Mitigation Project
KBBI	Kamus Besar Bahasa Indonesia (Great Dictionary of the Indonesian Language)
KKB	Kampung Kebon Bawang
KKM	Kampung Kamal Muara
KKP	Kementerian Kelautan dan Perikanan (Ministry of Ocean and Fisheries)
KMB	Kampung Muara Baru

LEAP	Locally embedded adaptation planning
LECZ	Low-elevation coastal zone
LRAP	Local Resilience Action Plan
MCI	Mercy Corps Indonesia
MDG	Millennium Development Goals
MIT	Massachusetts Institute of Technology
MOF	Ministry of Ocean and Fisheries
NCICD	National Capital Integrated Coastal Development
NGO	Nongovernment organization
OECF/JIBC	Overseas Economic Cooperation Fund
PAR	Pressures and Release
Perda	Peraturan Daerah (Regional Regulation)
PICAS	Planning for Integrated Coastal Adaptation Strategies
PIK	Pantai Indah Kapuk
PRA	Participatory Rapid Assessment
PT	Perseroan Terbatas (Limited Corporation)
RA	Risk Assessment
RAD-API	Rencana Aksi Daerah-Adaptasi Perubahan Iklim (adaptation planning for climate change at local level)
Raskin	Beras miskin (Rice for the poor)
RPB	Rencana Penanggulangan Bencana (Disaster Management Planning)
RPJMD	Rencana Pembangunan Jangka Menengah Daerah (Local Midterm Development Planning)
RT	Rukun Tetangga (Neighborhood Association)
RTRW	Rencana Tata Ruang Wilayah (general spatial planning)
RW	Rukun Warga (association of RTs, it represents the administrative unit of *kampung*)
SATLINMAS PBP	Satuan Perlindungan Masyarakat Penanggulangan Bencana dan Pengungsi (Government Protection Unit for Disaster Management and Refugees)
SLR	Sea level rise
SOP	Standard operating procedure
SREX	Special Report of the Intergovernmental Panel on Climate Change
START	Global Change System for Analysis, Research, and Training
Tagana	Taruna Siaga Bencana (Youth Organization for Disaster Response)
UKNA	Urban Knowledge Network Asia
UN DESA	United Nations Department of Economic and Social Affairs
UN-Habitat	United Nations Human Settlements Programme
UNDP	United Nations Development Programme
UNESCAP	United Nations Economic and Social Commission for Asia and the Pacific

UNESCO	United Nations Educational, Scientific and Cultural Organization
UNFCCC	United Nations Framework Convention on Climate Change
UNISDR	United Nations International Strategy for Disaster Reduction
UNU-EHS	United Nations University for Environment and Human Security
USAID	United States Agency for International Development
VCA	Vulnerability and Capacity Assessment
ZEF	Zentrum für Entwicklungsforschung (Center for Development Study)

List of Figures

Fig. 1.1	Asian cities at risk from sea level rise	3
Fig. 1.2	Focus area of study	9
Fig. 2.1	Illustration of *lifeworld* concept	22
Fig. 2.2	Vulnerability and vulnerable people	29
Fig. 2.3	People-centered adaptation planning framework	37
Fig. 3.1	Number of floods in Jakarta	40
Fig. 3.2	Observed rainfall in the period 1980–2012	41
Fig. 3.3	Illustration of tidal floods	42
Fig. 3.4	The distribution of poor households in Jakarta	44
Fig. 3.5	Flood areas due to rising sea level	49
Fig. 3.6	Shifted planning concept of flood control	52
Fig. 3.7	Illustration of the future vision of North Jakarta development	56
Fig. 3.8	The concept note of Jakarta Emergency Dredging Initiative (JEDI) in 2007	57
Fig. 3.9	Partnership by region	57
Fig. 3.10	The mainstreaming process of adaptation planning from community to city level	64
Fig. 4.1	The result of superimposed analysis	67
Fig. 4.2	Growth of KMB population	70
Fig. 4.3	Distribution of poor households	72
Fig. 4.4	The livelihoods of KMB residents	73
Fig. 4.5	Map of the KMB settlement	77
Fig. 4.6	The story of the informal sector	80
Fig. 4.7	The Interplay between poor settlements and floods	82
Fig. 4.8	Spatial boundaries of the Kampung Muara Baru people	83
Fig. 4.9	Temporal arrangements of floods in Kampung Muara Baru	85
Fig. 4.10	Flooding of KMB on 17 January 2013	85
Fig. 4.11	Fragile sea dyke in KMB. (a) Eroded. (b) Perforated	86

Fig. 4.12	Intentional decisions to live in KMB	88
Fig. 4.13	Javanese calendar related to rain	90
Fig. 4.14	Flood Map in 1992 and 2012 as perceived by the neighborhood leaders	95
Fig. 4.15	Adaptive capacity of Agus, the head of RT 15	100
Fig. 4.16	Vulnerability assessment using lifeworld analysis	101
Fig. 5.1	Model of triple-loop learning	106
Fig. 5.2	Shelter and evacuation maps	111
Fig. 5.3	Distribution of a two-floor House	112
Fig. 5.4	Adapted one-floor houses	113
Fig. 5.5	Adaptation plan for one-floor house	114
Fig. 5.6	Adapted two-floor houses; photo taken by the author	115
Fig. 5.7	Adaptation plan for a two-floor house	116
Fig. 5.8	Adapted settlement infrastructure, photo by the author and RW 17 collection	118
Fig. 5.9	Learning loop of KMB people	123
Fig. 5.10	Examples of activities of individuals that have generated income during the flood	125
Fig. 5.11	Transformation process from vulnerable to adaptation planners	128
Fig. 5.12	Staying on the roof	133
Fig. 5.13	The position of Rembuk RW in the planning system of Jakarta	137
Fig. 5.14	Rembuk RW 17 on 11 January 2013, photo taken by the Author	138
Fig. 5.15	Form of *Rembuk RW*	139
Fig. 5.16	The "back door" of the rock gate	140
Fig. 6.1	LEAP process (compiled by the author)	149

List of Tables

Table 2.1	The diversity of planning theory	18
Table 2.2	Identification of vulnerable people in vulnerability frameworks	33
Table 2.3	Approaches to vulnerability	34
Table 2.4	The difference between place-based and people-centered approaches	35
Table 3.1	Rainfall during two major floods of Jakarta	40
Table 3.2	History of flood management planning in Jakarta	45
Table 3.3	News regarding flood Jakarta	47
Table 3.4	History of coastal reclamation planning	48
Table 3.5	Planning related to flood of government of DKI Jakarta province	51
Table 3.6	Typology of processes of adaptation planning in Jakarta	62
Table 4.1	Number and types of buildings in Penjaringan	68
Table 4.2	Perceptions of the floods	89
Table 4.3	Lifeworld analysis of flood-related vulnerability	103
Table 5.1	Scenario planning for flood evacuation	108
Table 5.2	The multilevel institutionalization process	135
Table 6.1	Typology of adaptation planning	151

Chapter 1
Adaptation Planning for Floods

> *People may live differently in this world, but they do not live in different worlds*
>
> Antweiler 2012: 81

Abstract Increasing flood risks have demanded the emergence of planning theories and practices that can deal with the issue of environmental changes, such as climate-related disasters. This chapter portrays the urgency of floods to coastal cities in Southeast Asian region and expanding responses of flood-affected people to lowering their level of vulnerability. Kampung, where most of the urban poor lives, is a laboratory for discovering the embedded planning knowledge. The richness and diversity of adaptive practices are the potential sources of planning knowledge not only for local community but also for cities. Therefore, the people scale adaptation planning is emerged to provide an alternative solution in addressing flood problems.

Keywords Flood • Risk • Kampung • Poor • Coastal

Increasing flood risks have demanded the emergence of planning theories and practices that can deal with the issue of environmental changes, such as climate-related disasters. This chapter portrays the urgency of floods to coastal cities in Southeast Asian region and expanding responses of flood-affected people to lowering their level of vulnerability. Kampung, where most of the urban poor lives, is a laboratory for discovering the embedded planning knowledge. The richness and diversity of adaptive practices are the potential sources of planning knowledge not only for local community but also for cities. Therefore, the people scale adaptation planning is emerged to provide an alternative solution in addressing flood problems

1.1 Flooding as a Continuous Problem in Coastal Cities

Flooding is the most typical problem in urban development, especially in cities that lie in low-elevation coastal zones (LECZs). The natural characteristics of LECZs have exposed these coastal cities to ocean phenomena such as rising sea levels, high tides, and storm surges. Although they occupy only 2% of the world's land, coastal cities are the habitats of at least 13% of the world's population or about 60% of those who live in urban areas (McGranahan et al. 2007), including their assets and other supporting infrastructures. A recent study showed that the estimated losses caused by floods in the future would be much higher than the losses of today. In 2005, the average global flood losses were estimated to be approximately US $6 billion. Based on socioeconomic projections, this figure will increase to US $53 billion by 2050 or more than US $1 trillion per year if climate change and subsidence factors are added to the projection (Hallegatte et al. 2013).

In the last few decades, flood risk has increased because of fast-growing urbanization, climate change, and land subsidence (Hallegatte et al. 2013). Potential risks related to climate change are heightened by the accelerated rise in sea levels; increased sea surface temperatures; intensified extreme events, such as cyclones, extreme waves, and storm surges; altered precipitation patterns; and runoff (Nicholls et al. 2007). In 2007, the Intergovernmental Panel on Climate Change (IPCC) confirmed that several hotspots were situated in major Asian cities located near coastlines, rivers, and deltas, which indicated that their population and assets were at risk. Fuchs (2010) predicted that the Asian urban population would increase by 140,000 per day. He also predicted that "it will double from 1.25 billion in 2006 to 2.4 billion in 2030," and much of this growth will take place in coastal zones (see Fig. 1.1).

The severity of floods in Southeast Asian cities has been extreme, and in the last 2 years, three big floods have attracted the world's attention. First, the Bangkok flood in October 2011 was declared the worst in the last few decades, and it was covered by British Broadcasting Corporation (BBC) news for 2 consecutive weeks. This flood, which was triggered by heavy monsoon rains that began in July, inundated one-third of Thailand's provinces, including greater Bangkok, where manufacturing and other economic services were located. A local analyst estimated that the loss could account for 41% of Thailand's GDP (Harvey 2011). The second was the Vietnam flood, particularly in Quang Nam Province in November 2011, which caused the death of over 100 people (BBC 2011). The third was the Jakarta flood in January 2013, which was also covered by global media. It was reported that 15 people were killed, and 19,000 people were evacuated to shelters. The unusually heavy monsoon rains were also blamed as the main causal factor in the Jakarta flood (Quiano and Mullen 2013). The huge losses and damage caused by this flood were related to climate change and raised the awareness of the inhabitants of coastal cities regarding the importance of preparing for facing extreme flooding. In Southeast Asia, climate change is defined as a "nontraditional security issue" under the ASEAN Political-Security Community (APSC) Blueprint 2009–2015, which calls

1.1 Flooding as a Continuous Problem in Coastal Cities

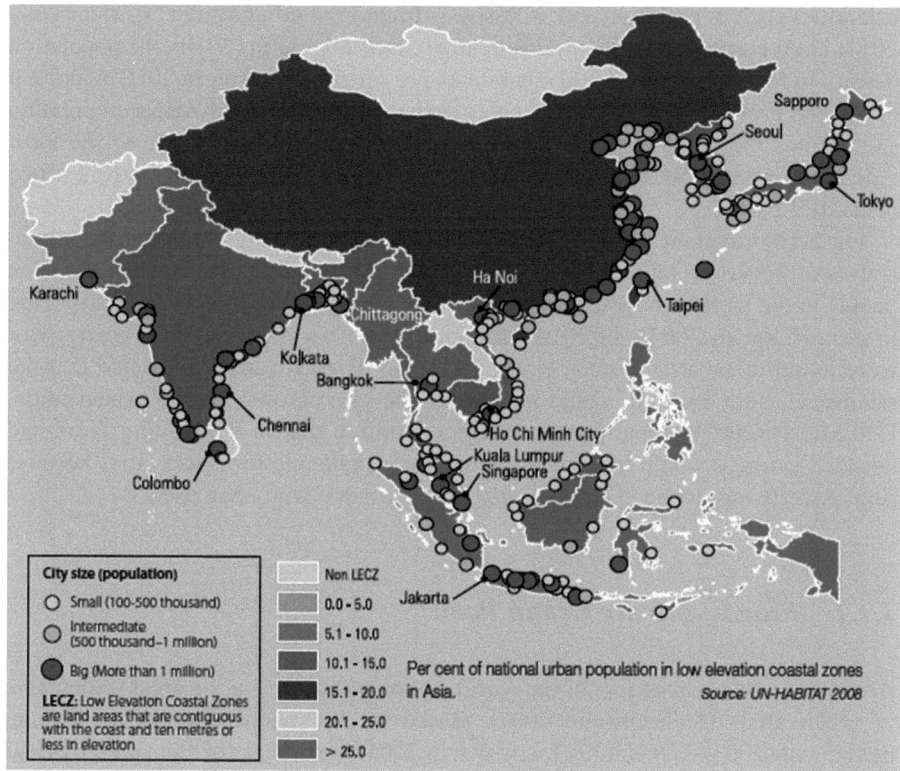

Fig. 1.1 Asian cities at risk from sea level rise (Source: Fuchs (2010))

for strengthening the cooperation under the ASEAN Agreement on Disaster Management and Emergency Response (AADMER) (Lian and Bhullar 2011).

Based on the cases of these coastal cities, the present changes to the waterways and drainage are not adequate to collect and channel huge volumes of water from torrential monsoon floods. Jamil and Ali (2013) noted that most plans in Southeast Asian cities tend to focus on immediate disaster relief, but they still face hurdles in implementing the plans at the local level and "in addressing the specific needs of informal settlement" (1). They pointed out that Jakarta and Manila have improper waste disposal systems, which lead to the clogged drains that exacerbate the effects of flooding.

North Coastal Jakarta,[1] where this research was conducted, "is particularly vulnerable to floods, rising seawater and other natural disasters as well as man-made

---

[1] North Jakarta City is located in the coastal area, which consists of six subdistricts: Penjaringan, Pademangan, Tanjung Priok, Koja, Kelapa Gading, and Cilincing. The land area is about 600 km², and the sea area is approximately 514 km². The sea area is part of Jakarta Bay, a shallow sea where the depth is generally not more than 30 m. The area is also a place for the end-flow of three main rivers (Citarum, Ciliwung, and Bekasi) and ten secondary rivers (Damar 2003). Therefore, North

calamities such as pollution and excessive extraction of groundwater" (Firman et al. 2011, 1). According to the Central Bureau of Statistics (BPS) (2010), the population density in Jakarta reached 1300 inhabitants per square kilometer in 2010, making it one of the most densely populated cities in the world. The highly dense population exacerbates the flood-related vulnerability of Jakarta, making it one of the most vulnerable cities in Southeast Asia (Yusuf and Fransisco 2009; Fuchs 2010; Firman et al. 2011).

Furthermore, many Asian cities currently are experiencing a trend of the urbanization of poverty (Ooi and Phua 2007). In this third millennium, of the 1.2 billion people in the world who are classified as poor, about 800 million or 60% are living in Asia (Laquian 2004). In 2010, around 400,000 poor and 300,000 nearly poor were living in Jakarta (Firman et al. 2011). UN Habitat (2010) predicted that the number of slum dwellers would increase at the same growth rate as urbanization. UNESCAP (2007) brought to the world's attention that in the coming years and decades, poverty would become a major urban challenge for Asian cities because rural poverty has declined significantly, while urban poverty has risen.

1.1.1 Floods and the Urban Poor

Around one in seven people of the world's population lives in informal settlements in urban areas (Satterthwaite 2007). Based on the Global Report on Human Settlement (2003), UN Habitat found that in 2001, the number of slum dwellers had reached 924 million, of which 28% were located in Southeast Asian regions. It estimated that by 2030, the global number of slum dwellers would reach 2 billion if no proper action was taken (UN Habitat 2013). A study of the geography of poverty, disasters, and climate extremes by Overseas Development Institute (ODI) reported that in 2030, "without concerted action, there could be up to 325 million extremely poor people living in the 49 countries most exposed to the full extent of natural disasters and climate extremes, in particular to droughts and floods" (ODI 2013, 7).

The IPCC has identified that urban informal settlements[2] are the most vulnerable to climate change (Watson et al. 1996) because many of them are built on hazardous

---

Jakarta is recognized as an area of coastal and delta cities as well. The engine of North Jakarta is the worldwide services of the sea harbor Tanjung Priok, supported by the industrial activities of multinational companies and the related activities, such as logistics, trading, offices, fisheries, and other services. About 48% of its area is used as residential area and 16% for industrial zone (Jakarta 2013). The future of this coastal city still relies on those base economic activities and is planned to expand through developing 12 reclaimed islands in the −8 m sea depth across the coastal area. The reclamation area is planned to be the place where the new brand of Jakarta as a global city will develop.

[2] According to UNSTAT as cited in Hofmann et al. (2008), an informal settlement is an area where "groups of housing units have been constructed on land that the occupants have no legal claim to, or occupy illegally; and/or an unplanned settlement and area where housing is not in compliance with current planning and building regulations (unauthorized housing)" (532). Therefore, informal

1.1 Flooding as a Continuous Problem in Coastal Cities

sites inhabited by a predominance of poor and migrant dwellers. Several scholars also pointed out that the impact of climate change has exacerbated existing urban settlements (Alam and Murry 2005; Banks et al. 2011), including informal settlements (CUS 2006). Others also argued that the poor living in various informal settlements are likely to be the most vulnerable because of natural hazards, which are driven by climate change (Moser and Satterthwaite 2009; Satterthwaite et al. 2009; Adger et al. 2003; McGranahan et al. 2007). Furthermore, Bosher et al. (2007) argued that key factors relating to poverty, marginalization, and powerlessness have shaped the severity of the effects of climate change. Thus, the presence of informal settlements is used to examine the effects of floods on the urban poor and vice versa.

UN Habitat (2003) argued that the failure of planning instruments (e.g., housing policies and spatial planning) was the major reason for the expansion of informal settlements in flood-prone areas, which are becoming visible. The World Bank (2011) affirmed that the lack of settlement capacity has caused the vulnerability of the urban poor. The absence of functioning storm drains, poorly built houses, and their location on the floodplain area are some of the reasons cited (Satterthwaite 2007). Urban institutions are usually unable or unwilling to address urban problems, such as floods in informal settlements, because informal settlers have no legal tenancy (Satterthwaite et al. 2009).

Recently, it has been recognized that the involvement of the urban poor should be increased instead of continuing to regard them as objects to be protected. One example is the initiative taken by the Asian Coalition for Community Action (ACCA) in 2009 across 19 Asian nations. This initiative used the model of "community-driven slum upgrading," in which donor grants support the initiatives of low-income communities to choose their own improvements (IIED 2012). The model is based on the view of urban poor as the key agents of change. The model places the urban poor at the center of the adaptive capacity of the city. Hence, the presence of the urban poor or slum dwellers is also important in making cities resilient to floods. The contribution of the urban poor in reducing urban vulnerability and increasing the adaptive capacity of cities cannot be overlooked. Therefore, how the urban poor manage floods in a process of environmental change is a key issue in urban development, particularly in Southeast Asian cities.

Given their limited resources, the urban poor usually count on their strong social networks, kinship ties, and active internal leadership (World Bank 2011) in dealing with floods. The practices of an urban community in adapting to floods (Roy et al. 2012) and existing coping strategies at the household level, including those of the poor (UNFCCC 2010), can be taken into account to optimize the adaptive capacity of cities. I argue that the adaptation plan can be built upon both practices. Therefore, the primary focus of this study concerns the question of how and what type of planning knowledge is produced from the perspective of urban poor.

---

settlement in this research refers to "the dense settlement encompassing poor communities housed in self-constructed shelters and living on public or private land without authorization" (532).

1.1.2 Kampung *as the Study's Focus*

According to Winayanti and Lang (2004, 42), "in Jakarta, the urban poor occupy a large number of spontaneous informal settlements referred to as *kampungs*." Because the buildings are irregular and the basic infrastructures are poor, the word *kampung* is usually attached to the word *kumuh* or slum. Moreover, if a *kampung kumuh* occupies an illegal plot of land (e.g., disputed land, state land, or private abandoned land), it is defined as *kampung liar* or squatter (Suparlan 2004; Durrand-Laserve 1998). In addition to that meaning, "the national government defines *kampung kumuh* as irregular settlements with substandard infrastructure, small plots of land for each housing unit, low quality of building structure and materials and illegal construction" (Silas 1990, 19).

There are about 600 *kampungs* in Jakarta (*Kompas*, 19 February 2000, cited in Sihombing 2010). Moreover, there are 490 pockets of poverty in Jakarta, which are always connected with the existence of *kampungs* (UN Habitat, 2003). "It is estimated that 20 to 25 percent of Jakarta residents live in *kampungs*, with an additional 4–5% squatting illegally along riverbanks, empty lots and floodplains" (UN Habitat 2003). However, because recent documentation is lacking, there are no recent figures for the exact number of *kampungs* or the actual *kampung* population in Jakarta (a census of *kampungs* has never been completed). Some scholars argued that the number of *kampungs* began to increase in the 1980s. About "85 percent of annual housing production in the 1980s…[was] developed by the occupants or residents themselves" (Struyk et al. 1990, 69), but there is no record of whether it was built informally or formally (based on government permits).

Kampung and *kota* (city) are two words that cannot be separated in describing the urbanization of Jakarta. In fact, Jakarta is "a result of territorialization of informal and formal land rights without the presence of effective state governance over the rapid urbanization" (Zhu and Simarmata 2015, 63). Although *kampung* and *kota* can be understood as conflicting images, there is a strong interdependency between them (Sihombing 2010). Wiryomartono used the notion of "*kampung–kota*" or "urban *kampung*" to describe "a settlement developed in an urban area without [basic] infrastructure planning or urban economic network" (Wiryomartono 1995, 171). Sihombing (2010) argued, "'*kampung-kota*' shows interdependency in social cultural, social economic, and social political terms." Thus, he argued that the word should be "'*kampungkota*,' which explains more closely the essence of the interdependency" (313). Therefore, discussions about Jakarta cannot ignore the presence of *kampungs*.

As an Indonesian urban planner, I am interested in examining the role of planning in providing safer locations for *kampungs* that lie in flood-prone areas. I am convinced that the urban planning process is imperative to ensure that the development pathway can increase the resiliency of cities. I have followed the work of several international scholars, such as Jöern Birkmann and Fernando (2008), who deal with the issue of spatial planning and disasters; Hans Füssel (2007), who deals with the issues of adaptation to climate change and planning; and Elizabeth Wilson

(2007), whose criticism of the planning profession addresses the issue of adaptation to climate change. In line with their works, I agree that the issues of climate change, especially climate-related disasters, have influenced the discourse of urban planning methods. The issues of scale, actors, and temporal dimension in developing a model of adaptation planning in particular are continuously evolving.

Why do I narrow my study to the community level, particularly to the *kampung*? First, because common features of most Indonesian cities were grounded in and formed by *kampungs*, urban development should focus on managing them. In the middle of mainstreamed globalized urbanization that imposes a new face of urban spaces in Indonesia (Bunnell et al. 2002; Kusno 2000), *kampung* still survived and to some extent evolved to provide urban services to many low-income groups of citizens. This self-mechanism settlement presents not to compete the property mainstream flows but fill the housing market for unskilled migrant labors, the poor, and low-income people. Even though it occurs in different places, I argue that the growing kampung formation performs an in situ urbanization.[3] It means that the urbanization process takes place in its original place, without land bank instrument.

The former governor of *Daerah Khusus Ibukota* (DKI) Jakarta province, Joko Widodo (the current President of the Republic of Indonesia), once stated that "*membangun Jakarta dimulai dari kampung*" (developing Jakarta started with the informal settlement) (Kompas 2013). He planned to improve the quality of *kampungs* at the rate of about 100 units per year (Rulistia 2012). From that standpoint, I suggest that those responsible for urban adaptation should be contextualized in the realm of the urban poor who live in *kampungs*, and the adaptation planning for floods should be built upon the planning knowledge that is embedded in *kampungs*. Therefore, with regard to the root of urban planning theory, I stand with the progressive mainstream rather than utopian visionaries. A new direction of the progressive movement could "help social structures learn from their experiences (social learning theories)" and seek "incremental changes that in the course of time would result in structural changes promoting equality, participation, and legitimacy" (Stiftel 2000, 9). As Indonesian cities are growing and expanding, the presence of kampung as the result of in situ urbanization needs to be studied deeper to understand the pathway. What I tried to build then is to develop a trackpath of how progressive planning can incorporate this social structure and accommodate various little chasing change of urban settlement to develop better planning praxis.

Compared to what the other planning scholars have done in the progressive way, this planning model emphasizes on understanding the level of individual knowledge on adaptation planning and following the transmitting process of the adaptation planning knowledge in the community. It differs from Howe and Langdon (2002) who examined the changing process more centralized to cultural capital; thus the social position in the community becomes the key driver. The reflective process that

---

[3] The term of in situ urbanization is similar with what occurs in China region in the case of transforming population, and settlement is without changing or moving out from the residential location, not the drivers of change.

plays an important role in planning is also based on the generalization of living space and life culture of the local people. I take a different angle by looking at the individual self-reflective practices. My focus is also different to what Hillier (2003) critics on Habermas' consensus space in planning process. Her interesting point of view on the role of politic agonism that breaks down the consensus building driven by Lacanian perspective is different to what lifeworld perspective I used that focuses on inner experiences of individuals that intershared each other. Regarding the planning context, I support what Aassche (2007) argued about the need of reflexivity when analyzing the context because the construction of context depends on "the framing of places by various stakeholders" (Aassche 2007, 114).

The second reason is that informal urban settlements, such as *kampungs*, provide interesting phenomena for dealing with the flood. I have witnessed that many informal settlements in Jakarta have occupied flood-prone areas for a long time. To date, the number of residents has increased, and the informal settlement areas have continuously expanded. The residents are still willing to stay in these areas without showing any intention to move to a safer place. In October 2013 at an international workshop[4] in Manchester, I discussed similar phenomena with the scholars, Anirudh Krishna, Regina John, and Afroza Parvin from global southern cities such as Khulna (Bangladesh), Bangalore (India), and Dar es Salaam (Tanzania). They agreed that the urban poor have their own coping strategies, which help them survive and accept the loss and damage caused by floods. Although floods have inundated their homeland for decades, they refuse to migrate. Therefore, the adaptation of the urban poor community is important to understand in order to know how people may live in ways that counter rationally unacceptable conditions.

In my research, I focus on the survival ability of slum dwellers in order to understand how they develop adaptation pathways under the threat of flooding. I use the term of adaptation pathway as an analytical approach to planning of urban poor community. I am convinced that the urban poor have their own mechanisms to deal with floods, and I consider that informal urban settlements provide the solution to the problem of making cities resilient. Because Jakarta city recognizes and admits the norm of survival, kampung is important to be internalized in the urban adaptation pathway.

Leaving aside the urban planning discipline for the moment and focusing on the community level, I build my study on the discipline of sociology, particularly microsociology, which concerns the nature of social interaction on a microscale. This approach will help urban planning to understand the context and the operational zones of planning practices. The outcome if planning considers the sociological approach is an increasing sense of belonging to the planning products. I examine the individual characteristics of *kampung* people and the social interactions that exist in *kampungs* (see Fig. 1.2), particularly on their pratices to adaptation planning.

---

[4] I was selected to present a paper at the International Workshop on "Living in Low-income Urban Settlements in an Era of Climate Change: Processes, Practices, Policies, and Politics," which was held by the Brooks World Poverty Institute (BWPI), University of Manchester in the Chancellors Hotel, 9–10 September 2013.

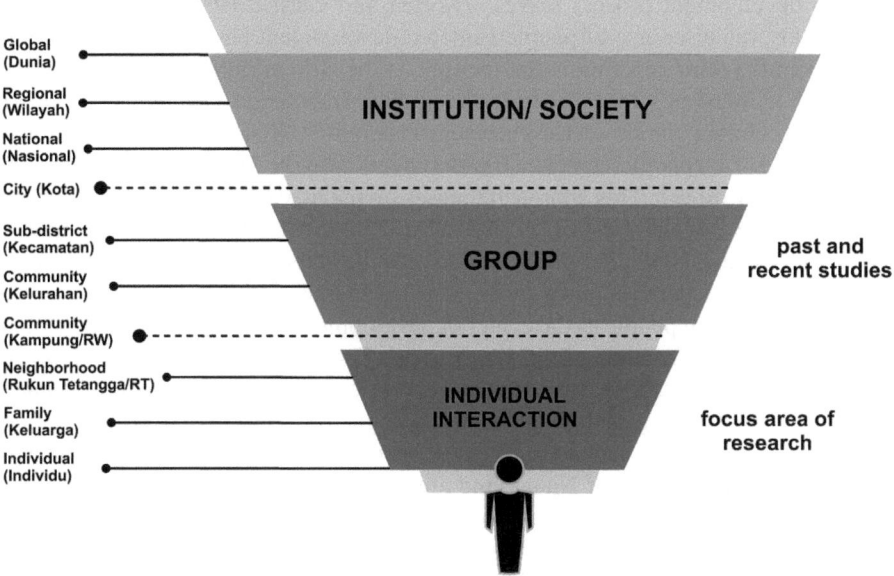

Fig. 1.2 Focus area of study

Despite Goffman's acting interaction, I use Schütz's lifeworld that pays attention on the consciousness of individuals in interacting with each other.

My research also extends to phenomenological sociology because it examines the structure of the lifeworld of *kampung* people. As explained before, kampung–kota, e.g., Jakarta, is quite different with kampung in rural area and probably in other Southeast Asian region, because it is populated by urban migrants that come from various islands of Indonesia. As Silas (1993) said, it is an incrementally developed self-help housing. And more than that, the rule of thumb in kampung–kota is also developed by the first generation in a casual manner. The rule can change depending on their consensus with seeing the city regulation as their concern. As noted by Ritzer (2011), the lifeworld "is an intersubjective world in which people both create social reality and are constrained by the preexisting social and cultural structures created by their predecessors." Therefore, the *kampung* is a socially constructed product.

1.2 People as Subjects of Adaptation to Climate Change

In development studies, positioning the urban poor as the subject of development is not new and has continuously evolved. Originating in David Korten's notion of people-centered development, this approach "looks to the creative initiative of people as the primary resource and to their material and spiritual well-being as the end that the development process serves" (Korten and Garner 1984, 201). Since the early 2000s, the discourse of people-centered development has entered a transition phase from the third generation that focuses on the human rights to the fourth generation that integrates local, social, and global knowledge (Korten and Garner 1984). The United Nation of Development Program (UNDP) (2011) reported that the approach of "people-centered development" was promoted through its 129 UNDP country offices. This approach focuses on the assessment of the needs of the people by the people themselves and assists them in addressing those needs. According to the UNDP (2011, 2), "People must be at the center of human development, both as beneficiaries and as drivers, as individuals and in groups."

The emphasis on people as the center of development is also present in the global debate of the periodization of the Earth's age. The Nobel Laureate Paul Crutzen popularized the concept of the Anthropocene to explain the new epoch of human dominance in changing and shaping the ecological system of the Earth. Steffen et al. (2007) argued, "human activities have become so pervasive and profound that they rival the great forces of Nature and are pushing the Earth into planetary terra incognita" (614). Several scholars also have agreed that the phenomenon of climate change in recent years has been induced by human activities, which they named anthropogenic climate change (IPCC 2001). Moreover, the global movement to enhance the human dimension in the development context has been addressed by several international networks, such as the International Human Dimension Program on Global Environmental Change (IHDP), which has the tagline of promoting "human action to global environmental challenge" in order to consolidate the various adaptation initiatives that have been taken by many countries. In this perspective, the framework of adaptation should be located among the people who live and carry out many practices of adaptation in their everyday lives.

Fifteen years ago, at an event held by the UN International Decade for Natural Disaster Reduction, the shifting focus on the human dimension in climate change adaptation was declared for the first time. The UN pointed out the following:

> The approaches to adaptation should move away from the notion of "climate change victims" and support the development of capacities for adaptation by the people, instead of for the people. It also suggested that the approaches to adaptation must learn from experiences in dealing with risk in development, and recognize the highly-differentiated nature of adaptive capacity and not prescribe "one-size-fits-all" solutions. [Therefore], the efforts must be concentrated on removing barriers and disincentives to autonomous adaptation to promote locally owned capacity development [processes]. (Christoplos et al. 2009)

> Thus, I argue that adaptation efforts must reflect the needs of vulnerable people. The adaptation pathway should be based on the learning processes in which people have tried to or failed to adapt to climate change. Because those who have lived for decades are actually the real practitioners. To some extent, they have developed capabilities in adapting to flooding. Therefore, the role of flood-affected people is imperative to appreciate the meaning of vulnerability and their adaption to it.

The role of the urban poor as the subject of adaptation depends on the degree to which they can utilize their own potential and capacity. According to Adger et al. (2009), "the poor may still strive to retain vitality and viability to cope with the shocks and/or adapt to the stresses of climate change" (341). It is different for the wealthy and those who can continue to enhance or maintain their current standard of living. Some scholars also pointed out that the local adaptation actions are reactive and only carried out on a short-term basis, unlike national or regional plans that are anticipatory and involve the formulation of policies and programs (Bohle 2001; Burton et al. 2002). They argued that poverty and extreme climate events limit the capacity of the poor to plan their adaptation pathway. In some cases, both poverty and climate extremes may force them to migrate (Brown 2008) or force the government to relocate them.

However, the informal settlements in Jakarta and other global southern cities have exhibited phenomena that differ from those argued for by the scholars mentioned previously. Informal settlements expand, and the urban poor still live there without accepting the option of relocation. Thus, a further question is raised: How do the poor, who have become used to flooding and who have stayed for a long time, build their own adaptation pathways? The poor who have experienced floods over time should have valuable knowledge about how to survive in flood situations and to anticipate their future impacts.

This examination of the poor in building their adaptation pathway is expected to strengthen and enrich the role of local knowledge in the debate about the adaptation to climate change. Local knowledge has been increasingly recognized as the key factor in building the capacity of risk management at the local level. The term "local" varies from the community level to the subnational level. In this research, I focus on the community of the poor because they have firsthand experience of the flood responses. The effects of flooding influence their everyday lives although they lack the assets and resources to cope with or to adapt to them. I argue that practical and traditional knowledge is produced by and within the community, which enables it to adapt to or cope with environmental changes, including floods. The local experience is a reservoir from which knowledge about disaster risk management and adaptation to climate change can be drawn (Tapsell et al. 2010).

Localities that grow in the realm of people through various ways influence the production of local knowledge. According to Hornidge and Antweiler (2012, 11), "the environmental changes that are currently being experienced by people across Southeast Asia, have led to a constant increase in uncertainties, insecurities and a lack of safety." This will lead to a further discussion about how and to what degree different localities are affected and how local people cope with and adapt to changed conditions. The focus on the process of knowing has also emerged because "the knowledgeability is different for the different members of a culture, and it changes, as it is itself a social product" (Hornidge and Antweiler 2012, 62). The meaning of environmental changes may be "different or even rival forms of knowledge" for different individuals. It depends on the content, practical relevance, and centrality of the knowledge. Through a case study of Makassar in Indonesia, it was demonstrated that "there were striking differences between the perceptions of local residents on the one hand and the language and concepts employed in official planning brochures on the other" (Antweiler 2012, 80).

On the methodological level, the recognition of how people adapt to environmental changes includes the understanding of the localities that evolve and are institutionalized in the community. Asking for people's perceptions through questionnaires and interpreting their responses through descriptive statistical analysis will not obtain the essential meaning. A representation of one phenomenon in society can be achieved, but it fails to explicate the deep values that constitute the social world of a community, especially the poor whose lives are based on informality. This methodology leads to the misinterpretation of the needs of poor people who often covertly adapt because of their lack of resources and high dependency on informal institutions. Consequently, adaptation planning, which is driven by external actors, can be mistaken in its implementation in poor communities. As pointed out by Baum (1980), "most planners do not see the world in which they practice in either political or organizational terms" (296). It thus leads to faulty interpretations of what the poor need and how they plan for the adaptation to disasters such as floods.

I assume that the current process of adaptation planning involves only vulnerable people in order to confirm experts' assessments of their condition. The process has been conducted in the technical domain of planners. This methodology does not allow adequate time and space for vulnerable people to reveal their knowledge, assess their conditions, and plan their adaptation pathways. In this respect, I suggest that the starting point for building an adaptation pathway should be based on the realm and the needs of the poor. By doing so, the adaptation pathway could be customized to local circumstances and be implemented by local people.

1.3 People-Based Adaptation Planning

Centering the planning process on people has been discussed for decades. People have always adapted to the changes in their lives, and the adaptation creates new knowledge that helps them to be more resilient in uncertain times. Thus, the planning process should consider the individual actors, subjective cognitions, emotions, egoism, and the "subjective logic" of their argumentation (Weichhart 2010, 12). Many planning theorists, such as Friedmann (1969), Dalton (1989), Forester and Stitzel (1989), and Alexander (1992), believe that planning is not limited to and does not belong to planners only. Planning also belongs to people who can frame their own problems and propose reflective actions.

Several studies of community-based adaptation have attempted to develop a planning model that is based on people's participation. This model applies a certain planning procedure by using the assistance of external parties to facilitate the process, such as nongovernmental organizations (NGOs), professional associations, and local governments. The participatory planning model uses the theory of communicative action developed by Jürgen Habermas, a prominent German philosopher. He argued that communicative action "seeks to reach an understanding of the situation and their plans of action in order to coordinate their actions by way of agreement" (Habermas 1984, 86). Hence, by acting communicatively, planners should be able to establish a planning agenda, encourage public participation, and provide alternative solutions for communities. Planners can assist in ensuring that the rationalization process is successful. Therefore, this approach accommodates the voices of the poor and provides technical assistance in order to enrich the knowledge of the urban poor.

Accordingly, planners are predominant in framing the communication process, although they accommodate some input from the community. Planners assume that adaptation knowledge is a technical term that pertains to their professional domain. The role of planners as facilitators is to establish the planning process, which then necessitates a high degree of dependency on the presence of the planners. Several initiatives may have been accomplished successfully and consequently strengthened the community, but they may have led the people in a "new world" direction, which at times does not fit or reflect their reality.

Victoria Beard (2002) also studied how local people engage in the planning process. She found that planning embodies the desire for emancipation from local perceptions, which she called covert planning. It is likely that informal planning "involves local people planning for them and, in so doing, taking incremental, incipient steps toward altering larger power relation" (Beard 2002, 18). Informal planning transforms individual planning initiatives into community planning through a subtle and nuanced form of collective action.

Nonetheless, few studies have examined "planning that is embodied in the process of subjective human experiences" (Wagner 1970, 13). In terms of adaptation to climate change, little has been done to investigate planning practices in the community. As an urban planner, I am interested in understanding the planning process of the urban poor in adapting to environmental changes. They have demonstrated that they can live in flood-prone areas for decades. Many nonpoor people perceive this choice as irrational; however, the urban poor have managed to live continuously in these areas.

However, a planning model that is purposively designed to reveal shared realities in the social world has not yet been developed. The conception of *lifeworld* is seldom used to define planning processes that are rooted in the realm of people. Habermas used the conception of *lifeworld* as an object wherein communicative action takes place. Donald Schön (1983), another prominent philosopher, argued that planning could be proceeded by reflecting on actions or reflective practices. He argued that planning is a universal human activity that depends on "the capacity to reflect on action so as to engage in a process of continuous learning" (Schön 1983, 8). As I understand it, this capacity is related to the concept of *lifeworld* developed by Alfred Schütz, a prominent Austrian phenomenologist. According to Münch, "the *lifeworld* is an inter-subjectively shared world, a stock of knowledge, consisting of typifications, abilities, skills, and recipes for observing, interpreting the world, and acting in this world" (cited in Oberkircher and Hornidge 2011).

Schütz's concept of the individual *lifeworld* is in line with Schön's conception of the practitioner's capacity. According to Schön, practitioners "often reveal a capacity for reflection on their intuitive knowing in the midst of action…" (Schön 1983,

9). In a critic of the technical rationality approach, Schön argued that the planning problem needs artistry and a set of skills that go far beyond a theoretical base (McDowell et al. 2007). In terms of creating solutions, planning does not necessarily depend on espoused theories (i.e., research based) but on experiences (theory-in-use) that were embedded in the logic of the action. Therefore, "the reflective practice is about awareness of the knowledge we use, how we use it and how we can improve our action in real time" (McDowell et al. 2007, 10). Schön argued that "planning knowledge is analyzed as a system of knowing-in-practice that includes the framing of role and situation, and the interpersonal theories of action which the practitioner brings to his practice" (Schön 1983, 352).

Both Schön and Schütz used the argument that knowledge comes from repeated actions or practices by different actors. Schön defined practitioners as professionals and developed a model to increase their capacity (internal use). Schütz defined practitioners as ordinary people who actively practice (routinize) and share the meaning of the phenomenon to simplify everyday life (habitual knowledge) (Schütz 1973). However, Schütz's *lifeworld* has never been used to examine how planning operates in a community. Therefore, this research demonstrates that the *lifeworld* is a domain to increase the reflective capacity of people in making an adaptation plan. As the American Institute of Certified Planning (AICP) stated, planning is a universal human activity (AICP 2013) and can be derived simply from reflection in action (Schön 1987).

Adaptation planning belongs to not only professional planners and planning experts but also common people. Planning is a universal human activity that involves reflecting on past experiences (precedent world) and future expectations (subsequent world). The intention to create a planning model that can facilitate a socially reflective action has never been applied to investigate adaptation at the community level. Laukkonen et al. (2009) recommended the development of a methodology and tools to help individuals and communities in the planning process not only to encourage them to participate but also to embed their knowledge and actions in the adaptation planning process.

Vulnerable communities that have been frequently affected by floods know that they can adapt to severe conditions from time to time. They may not know that their reflections on past experiences and future actions are a form of planning knowledge. However, reflection could be used to develop a model of adaptation planning. Moreover, the planning knowledge that depends on local perceptions and institutions is already embedded in the community. We simply do not know what the process will be and how it will take place. Research is needed on how people plan in their everyday life, particularly the type of planning that exteriorizes local adaptation pathways and their way to institutionalize the plan.

In this study, the urban poor were selected because their planning habits have been underestimated in the discussion on adaptation to climate change. Most nonpoor people presume that the urban poor have no planning practices in their everyday lives and that planning knowledge belongs only to professional planners. Thus, the primary focus of this study concerns the question of how and what type of planning knowledge is produced from the perspective of urban poor.

At present, the research on adaptation planning in Jakarta has focused almost exclusively on integrating technical climate indicators or scenarios of planning processes that are performed by professional planners. At least four planning activities have been conducted at the city level: (1) the Detailed Spatial Planning of the Penjaringan Subdistrict; (2) the Replanning of North Coastal Jakarta by the Government of DKI Jakarta in 2011; and (3) Jakarta Climate Adaptation Tools and the Jakarta Coastal Development Strategy in 2011 conducted by the Royal Haskoning MSC, which was donated by the government of the Netherlands (Elings 2011). The fourth activity was the Jakarta Emergency Dredging Initiative (JEDI) and the Alliance of Green Delta City Defense Planning through major storm, water drainage, and canal systems, which was funded by the World Bank in 2009 (Prasad et al. 2009). At the community level, most of the initiatives attempted to combine technical and local knowledge. For example, the first initiative was the community-based disaster risk management, founded by *Action Contre La Faim* (ACF) in 2008. The second initiative was the Mercy Corps program, Stakeholder Coordination, Advocacy, Linkage, and Engagement for Resilience (SCALE) project, which was followed by the Asian Cities Climate Change Resilience Network (ACCCRN) program in 2011.

Furthermore, the participatory planning approach has been applied in the practice of adaptation planning at the community level. This planning process places the experts and the community on the same level, but it does not position people as main actors in the process. It often happens that the planned adaptation, as viewed only from the professionals' point of view, might not be congruent with perceptions of vulnerable people. Thus, a study based only on technical rationality would not uncover the essential meaning of adaptation planning as experienced by vulnerable people. We need to learn more about the "lived experiences" of individuals in adaptation planning and the meaning that they ascribe to their everyday lives.

This book raises the main question that asks how individuals, based on their *lifeworld*, practice and institutionalize adaptation planning in their community. To answer this question, I need to address first the context and the locus of the adaptation planning taking place. Thus, there are two sub-questions: What kinds of planning institutions are constituted in Jakarta? How do *kampungs* represent the interplay between the urban poor and regular flooding? The following questions focus on the *kampungs*; each question consists of three sub-questions. The first question concerns vulnerable people's perceptions of a flood: Who is vulnerable and why are they vulnerable? How do they define vulnerability? In addition, how do they define adapters? The second question is related to the meaning of planned adaptation: How do they practice adaptation? What are their experiences in planning adaptation? What meanings do they assign to those experiences? The third question concerns the institutionalization and reification process: How did they organize and implement the plan? How did they institutionalize it? Thus, I wish to examine adaptation planning based on their perceptions and meanings.

Therefore, the purpose of this study is to understand the institutionalization and reification process of adaptation planning for floods from the perspectives of individual *lifeworld*s in the local community through identifying vulnerable people who have experience in planned adaptation, examining the perception and meaning of flood-related vulnerability, and figuring out the process of the institutionalization and the reification of adaptation planning in this social world. I examine how the urban poor acquire and incorporate reasonable understanding of their adaptation for floods. Despite the social network, I prefer to examine the repetitive or changed patterns of their "habitualized activities" in adapting their lives to the threat of floods and the adaptive actions that help them to cope automatically with repeated flood situations. Because they live in the same place, I presume that they have the opportunity to observe and respond to each other's practices and to anticipate the habits of others. Over time, these responses and habits are institutionalized, enacted, and controlled until it is forgotten that such institutions were human-made.

This study is aligned with and supports the emergence of the people-based adaptation planning. It is needed because first, adaptation planning cannot depend only on climatic proofing because in practice, climatic data is still lacking in terms of types and depth of scale, and climate modeling is still weak in providing accurate future scenarios, which makes it difficult to conduct meaningful assessments of the effects of climate (Firman et al. 2011). Second, participatory planning has shown weaknesses in providing the space and time required to investigate the endogenous and intentional motives of local people and to allow them to articulate and decide their own actions. The agenda setting of third parties (e.g., NGOs, donors, or other organizations) is inevitable. Third, the integration of local and scientific knowledge in adaptation planning requires synchronizing tools to connect the different experiences of local people and scientific experts who derive variables from their understanding. Fourth, as suggested by Antweiler (2004), the citizens in Indonesia could be positioned as experts in generating urban knowledge. Local knowledge is "a locally situated form of knowledge and performance which comprises skills and acquired intelligence responding to constant social and environmental changes" (Antweiler 2004, 1). Therefore, a localized approach to creating adaptation planning is necessary.

Chapter 2
Lifeworld as the Domain of Adaptation Planning

Abstract Based on Schütz's *lifeworld* and Schön's reflective practices, I intend to examine the adaptation planning process of the urban poor. The process is personal and tacit because they live in informal settlements. A planning framework that reflects the past experiences (precedent world) and future expectations (subsequent world) of the urban poor uses the geographical setting of the urban poor, who have practiced adaptation in their everyday lives, and it utilizes locally embedded knowledge. Knowing the *lifeworld* of flood-affected people is an important step to initiate the adaptation planning process.

Keywords Lifeworld • Reflective practice • Phenomenology • Planning • Knowledge

Based on Schütz's *lifeworld* and Schön's reflective practices, I intend to examine the adaptation planning process of the urban poor. The process is personal and tacit because they live in informal settlements. A planning framework that reflects the past experiences (precedent world) and future expectations (subsequent world) of the urban poor uses the geographical setting of the urban poor, who have practiced adaptation in their everyday lives, and it utilizes locally embedded knowledge. Knowing the *lifeworld* of flood-affected people is an important step to initiate the adaptation planning process.

2.1 Lifeworld and Planning

Planning theories and practices cannot be separated from the social realm. Based on the history of planning in the USA, social science emerged in the era of the Great Depression and urban stagnation. Since the 1960s, the rise of social activism has both strengthened and challenged social science. The increase in the number of community voices and social protests, in addition to political actions for reformation and transformation, has dominated the discourse of planning. Since the 1980s, in the era of retreat and policy privatization, the postmodern critique of rationality has stressed interaction, communication, and process. It shifts the planning

Table 2.1 The diversity of planning theory

Linkage between knowledge and action	Desired outcome	
	System improving	System transforming
Cognitive reality	Synoptic rationality	Radical planning
Procedural rationality	Incrementalism comprehensive planning	Advocacy planning
Communicative rationality	Traditional participatory planning	Transactive/collaborative planning mediation
Self-reflective action	*Phenomenology*	Critical theory
	Social learning	Social mobilization
Moral philosophy	–	Utopianism

Source: American Institute of Certified Planners (AICP) (2013)

paradigm from the voices of communities to the communities with voices (AICP 2013). Therefore, the linkage between knowledge and action has defined planning outcomes. The American Institute of Certified Planners (AICP) divided planning into several branches of theory, which the present study takes into account (see Table 2.1). This book is positioned in the branch of phenomenology in planning.

Donald Schön (1987) introduced reflective practice to explain the relation between thought and action. This concept was developed to improve education in professional schools, including planning schools. Schön presented reflective practice as a response to technical rationality (McDowell et al. 2007). He argued that planning needs artistry and skills that go far beyond the theoretical base. It does not depend on espoused theory (research based) but on the experiences (theory in use) that are embedded in the logic of the action for creating solutions. Therefore, "reflective practice is about awareness of the knowledge we use, how that use is, and how we can improve our action in real time" (McDowell et al. 2007, 10). The reflection on knowing in action is the key to developing a planning process. In 1978, Aagrys and Schön proposed the learning loop framework, which could also be used to increase adaptive capacity (Pahl-Wost 2009; Flood and Romm 1996).

Planning theorists and practitioners, such as Friedmann (1987), Forester and Stitzel (1989), Alexander (1992), and (Fainstein and Campbell 2003), believed that the planning process is not exclusive to planners. Others also use this process to explain their problems and propose reflective solutions. Specifically, planning should consider individual actors, subjective cognitions, emotions, egoism, and the "subjective logic" of their argumentation (Weichhart 2010,12) so that the outcome serves the actors' interests (Briassoulis 1997). Therefore, planning is not expected to produce texts that do not consider the local context (Sudaryono 2012).

Covert planning focuses on how local people engage in the planning process. Proposed by Victoria Beard in 2002, it draws on Friedmann's theory of planning as social mobilization. Covert planning thus embodies the desire for emancipation from local perceptions. It is more likely to be an informal planning that "involves local people planning for them and, in so doing, taking incremental, incipient steps toward altering larger power relation" (Beard 2002). It transforms individual planning initiatives into community planning through a subtle and nuanced form of

2.1 Lifeworld and Planning

collective action. Nonetheless, few studies have examined the embodiment of planning in "the process of subjective human experiences" (Wagner 1970, 13) or its process in individual reflective practices, especially in relation to climate change adaptation.

As a new dimension of planning theories and practices, adaptation to climate change can be seen as a new problem that should be incorporated into the planning process. At regional and city levels, most adaptation plans work using the espoused theory. Although professional planners have incorporated adaptation into planning, they have not had to adapt it to a flood. They use climate modeling data and the technical advice of experts and resource persons who know about floods. Such data and information are elaborated to produce planning conceptions without considering people's participation.

However, at the community level, participation has been recognized as an important factor. Many initiatives have used participatory planning as the theoretical framework. This framework is used to explain communicative actions of the community and external factors, such as NGOs, the role of each actor, and the degree of participation in the adaptation planning process. However, this framework does not include the endogenous knowledge that may have accumulated in the realm of flood-affected people because NGOs have their own ways of orchestrating the planning process.

Where do the flood-affected people fit? They are the real experts in adapting to floods because they have done so for decades. Hence, they are knowledgeable. How do they frame the flood problem that they hope to solve? What steps do they enact when they plan the adaptation? A theory in use should be embedded in the logic of their adaption pathway. Therefore, the reflection on past experiences in adapting to floods is the main source of an adaptation plan. Based on that reflection, they create solutions and achieve capabilities in planning the adaption.

Phenomenology in planning is concerned with people's understanding of planning practices. It is derived from the criticism of planning practices that they are not located in and connected to the real world (Baum 1980). John Forester said that such practices would continue to separate planners and those for whom they plan. According to Forester (cited in Hummel 1982), three issues need to be addressed in the phenomenology of planning:

> A critical theory of planning practice calls our attention (1) empirically to concrete communicative actions and/organizational and political economic structures; (2) interpretively to the meanings and experiences of persons performing or facing those communicative actions; and (3) normatively to the respect for or violation of fundamental social norms of language use, norms making possible the very intelligibility and common sense of our social world. (Forester et al. 1980, 337)

Therefore, phenomenology in planning focuses on the intentionality of human acts. Its function is that of validation by placing "what is" in a context of meaning. It provides a domain for people's actions in making plans for their own reasons:

> Phenomenology is the technical bridging of subject-object, reason-practical reason and fact-value dichotomies that alienate human beings from their own potential (self-) knowl-

edge. Phenomenology becomes the tool of a determined effort to find new ways of knowing as a prelude to new ways of deciding and new ways of acting. (Denhardt 1981, 100)

In people's adaptation to floods, their motives, processes, and institutionalization are the main phenomenological concerns in planning.

Sudaryono (2012) argued that modern planning practices in Indonesia tend to "consider that the truth of procedural aspects has only produced the text, without context" (6). He suggested using "Husserl's lived experience" or "Schütz's *lifeworld*" to examine the planning process so that planners could observe a real world. Hence, the role of planners should shift from guardians of the truth who apply deductive positivism to truth diggers who apply inductive–explorative research (Sudaryono 2012, 10). Compared to Habermas's theory of communicative action, which focuses on differentiation between system and *lifeworld*, Sudaryono's approach will help the researcher to gain firsthand experience in the operation of the planning process.

Planning processes should consider contexts, multiple disciplines, and organizational knowledge. Planning can be effective if we "understand about an institution in general and know their specific institutional contexts in particular" (Alexander 2005, 210). Institutions can provide the context and framework in which planning operates (Verma 2007). However, the institutionalization of planning takes place in the social world where people are knowledgeable about their actions and interests. Teitz (2007) argued that the key issue is the aggregation of individual preferences and collective decisions. According to Healey (2007), institutions are a kind of "soft infrastructure of the governance of social life" (Healey 2007, 65). Furthermore, "Institutions structure the interactional process through which preferences and interests are articulated and decisions made" (64). Therefore, the outcomes always have difficulty in achieving social acceptance if the planning is not embedded in institutions. Legitimation is lacking when the theme of planning falls outside "norms, rules, and practices" (Giddens 1984; DiMaggio and Powell 1991, cited in Healey 2007). Several scholars have argued that informality has become a way of life in third-world countries (Alsayyad and Roy 2004; Hasan 2004; Bayat 2000; Moser 1978).

Briassoulis identified two dimensions of planning: formal and informal. "[F]ormal planning is crystallized in rules and norms received through academia and legitimized by state legislation" and "informal planning is a universal human activity, not limited to the public sector, and thus not necessarily (formally) institutionalized" (Briassoulis 1997, 105). The two types of planning are combined (Hodgson 2006). Therefore, planning does not depend on or consider formal and informal institutions. The human factor is essential to planning. If the needs of the people are not recognized or if they have no access to the formal planning process, they can use their own knowledge of the past to proceed with their own planning. How they do this is worthy of study.

The theory of communicative action proposes that *lifeworldly* communication enables planning to achieve mutual understanding. Habermas (1984) argued that understanding is possible if the representatives of interest groups are ready to relinquish power and initiate "an exploration mutual argumentation for basic under-

2.1 Lifeworld and Planning

standing and values upon which to build consensual planning decision" (Mäntysalo 2004, 10). In his theory, Habermas divides society into system and *lifeworld*. System is the realm of power and money, and *lifeworld* is the realm of cultural production and the reproduction of values (Mäntysalo 2004). Habermas placed media as a "subsystem" of *lifeworld* that emerged from the *lifeworld* and began to dominate society.

Mäntysalo (2004) argued that Habermas's idea met with some criticism, especially by Foucauldians who argued that power is not an "outer distortion" of the *lifeworld* but is embedded in and has deeper effects on the bureaucratization and commodification of society. Hillier (2003) argues that antagonism is the constitutive outside of consensus formation (53). Power is thus seen as a constructive force that shapes the individual's understanding and perception. However, for Habermas, the cultural and the social are included in the "power-free" *lifeworld*. Furthermore, the debate raises the issue of self-clarity in the institutionalization of planning. Following Fay's (1987) argument that self-clarity is problematic, Huxley (2000) argued that it is difficult because different customs and traditions are situated and embedded in the realm of individuals. Therefore, self-knowledge must be revealed.

Alfred Schütz (1899–1959) linked the phenomenology of Edmund Husserl to the sociology of Max Weber. In Schütz's *lifeworld*, planning is personal and based on inner experience. Schütz saw the *lifeworld* (*Lebenswelt*) as consisting of life forms and structures that refine meaning, action, and intersubjectivity. Münch explained, "*lifeworld* is inter-subjectively shared world, a stock of knowledge, consisting of typifications, abilities, important skills and recipes for observing, interpreting the world and acting in this world" (Münch 2003, cited in Oberkicher and Hornidge 2011, 398). It is a lived world of everyday life, a structure in which the provinces of realities with finite meaning-structures are consciously constructed from the lived experiences of individuals.

Spatial, social, and temporal arrangements constitute the lived world and delineate the zone of operation (*Wirkzone*) of individuals, which depends on the cumulative past, completed experiences, and relatively open expectations for the future (Adomßent 2004, cited in Oberkicher and Hornidge 2011, 10). Figure 2.1 shows that the individual's provinces of reality are structured by four worlds: "the individual's immediate environment (*Umwelt*), the surrounding world (*Mitwelt*), the world of predecessors or precedent world (*Vorwelt*), and the world of successors or subsequent world (*Folgewelt*)" (Schütz and Luckman 1974, 59). The everyday *lifeworld* is the "reality" of people. Furthermore, Schütz argued that the social world is the sum of many individual lifeworlds. In other words, a lifeworld is based on each person's social and cultural experiences and their associated meanings, which depend on and are determined by the person's position in time and space (Schütz 1973; Schütz and Luckmann 1974). Schütz's lifeworld is useful to understand the individual actions in term of adaptation rather than Goffman's acting interaction that assumes the social world is a stage (Goffman 1959). In the context of flood disaster, flood-affected people tends to act and react as the way they are since it will influence their life.

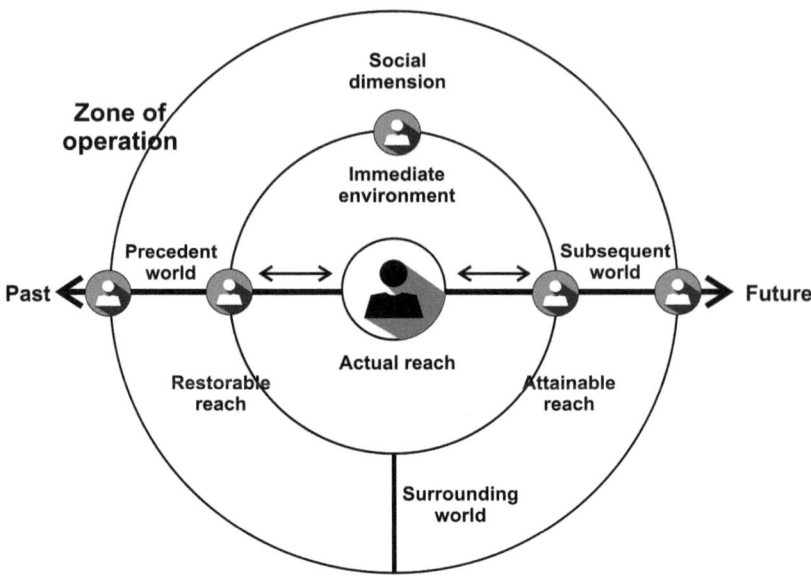

Fig. 2.1 Illustration of *lifeworld* concept (Source: Authors, based on Schütz and Luckmann (1974))

The social world can be defined through the socially constructed symbolic system developed by Berger and Luckmann in 1984. According to this system, there are three kinds of construction processes: externalization, objectification, and internalization. These processes take place through typification, institutionalization, legitimization, and reification. Typification consists of the shared habitual processes that lead us to typify and categorize what we observe in our surroundings through individual interactions. Institutionalization occurs if all members of a society share knowledge through a set of rules or legal institutions. Each rule will be accepted and passed on from generation to generations in processes of legitimization. When the institutions are no longer questioned, and they are accepted as social facts, the reification phase begins (Hornidge 2011). In reification, the objectivity of the social world has taken place, which then confronts humans as something outside themselves. Reification implies that "man is capable of forgetting his own authorship of the human world, and further, that the dialectic between man, the producer, and his products is lost to consciousness" (Berger and Luckmann 1967a, b, 89).

A *lifeworld* analysis offers a social reality that "is not replaced by a fictitious, non-existent world constructed by some scientific observer" (Flick et al. 2004, 69). In several case studies, the *lifeworld* produces meaning from individual experience, which later can be applied to planning. Meaning is "the product of active local interests and social communities and constitutive of their interests" (Heelan 2001, 7). It does not mean that meaning is without truth, but it is the place where truth makes meaning appear (Heelan 2001). For example, Veronica Strang revealed that "the broad themes of meanings encoded in water are similar in substance, providing important undercurrents of commonality" though culturally specific and diverse in form (Strang 2005, 115). The *lifeworld* has also contributed to explaining the "processes of perceiving and interpreting water and… [its] influence on water use based on farmer's perspectives in Khorezm, Uzbekistan" (Oberkicher and Hornidge 2011, 397). The meanings of water can actually tell us about community members' perceptions of the phenomenon.

Several scholars in *Zentrum für Entwicklungsforschung* (ZEF), where I work, have applied Schütz's *lifeworld* to the field of disaster risk studies and climate change adaptation. For example, they have revealed the practices of flood utilization for livelihoods of Vietnamese farmers in the Mekong delta region (Ehlert 2012), investigated the local knowledge applied in the adaptation strategies to cope with drought and flooding in Toineke Village in West Timor, Indonesia (Hornidge and Schotes 2012), and examined the role of religious values in the water management implemented by Khorezm's farmers (Oberkicher and Hornidge 2011). Therefore, the *lifeworld* is suited to examining the lived experiences of people with regard to a phenomenon.

In phenomenological sociology, Schütz's *lifeworld* provides an analytical explanation of intentional, conscious actions taken to discover social existence. It is used to examine the institutionalization of adaptation planning as a social construction because the *lifeworld* provides the context and the domain of adaptation planning processes. Schütz's *lifeworld* can also be used to explain the role of consciousness and actions of people in transforming subjective knowledge to the social stock of knowledge.

In the phenomenology of planning, Schütz's *lifeworld* can be used to operationalize Schön's reflective practices in the sociological dimension. For example, those who have practiced adaptation for decades could replace Schön's reflective practitioners. Moreover, the capacity of practitioners could be defined by the zone of operation of individuals in the shared subjective meaning of adaptation itself. Therefore, reflective practice happens when individuals reflect on their world and its surroundings to arrive at a plausible reason for projecting their future actions in adaptation. This will generate a stock of knowledge based on the experiences of the people.

As argued by Michael Barber, "in planning some actions to be recognized in the future, one trusts reflective acts of projection, like those found in reflective memory, only now oriented in a future as opposed to past direction" (Barber 2014, 4). Through such reflectivity, one can establish the "in-order-to motive of one's action." Therefore, this research requires a phenomenological empirical study to discover the structure of *lifeworld* of the *kampung* people who endure regular floods in order to interpret the meanings and experiences of those adapting to the flood situation and to understand their social world through institutionalization.

2.2 Adaptation Planning: Community- and People-Based Approaches

Adaptation to environmental change is as old as the human presence on Earth (Smithers and Smit 1997) although new needs have arisen in climate change adaptation. These needs include responding to changes and adapting successfully in the future. According to Lisa Schipper (2007, 4), "adaptation to climate change can be imposed based on premeditated planning or it can take place without specific policy framework to implement it." Therefore, the discourse on adaptation planning has entered the realm of sustainable urban development, and it has involved many development agencies.

In 2010, at Changwon, Korea, UN Habitat held a meeting of regional partners in Asia Pacific countries to discuss the "Cities in Climate Change Initiative." UN Habitat declared the importance of strengthening cities' responses to climate change, especially for vulnerable youth and women, in addition to strengthening responses to local climate change through the integration of community involvement, private sector participation, and changes in laws and regulations (UN Habitat 2010). These responses should be on the agendas of both cities and communities. However, there have been "few concerted efforts to develop dedicated adaptation plans or to set adaptation initiatives in motion" (Carmin et al. 2012).

Unlike climate change mitigation, international protocols have not involved adaptation planning procedures at the municipal level, so some cities have initiated their own adaptation plans (Carmin et al. 2012; Granberg and Elander 2007; Schreurs 2008) with the support of international donors. For instance, in 2009, the World Bank released the *Climate Change Adaptation Handbook* for the Mayor and the Rockefeller Foundation that applied resilience concepts to planning practices in the Asian Cities Climate Change Resilience Network (ACCCRN) program (2009–2011). The World Bank's adaptation planning process is similar to a risk reduction program (Thakoerdin 2009). ACCCRN identified four key elements of adaptation planning: exposure, systems, agents, and institutions. It recommends that the planning process should integrate local knowledge and scientific knowledge (Moench et al. 2011).

I argue that adaptation planning proceeded in three modes: first is focusing on sectoral issues, second is incorporating the adaptation into statutory planning, and third is conducting a specific purpose planning for adaptation. The first mode uses a strategic approach to accommodate sectoral needs, such as water management, poverty reduction, disaster preparedness, and city planning (Fussel 2007). Fussel believed that the benefit of adaptation planning is to fill "the gap between expected (assume only autonomous adaptation) and residual impacts (assume autonomous and feasible planned adaptation) rather than potential (no adaptation) unavoidable impacts (perfect adaptation)" (269). He argued that adaptation planning concerns "making recommendations about who should do what more, less, or differently, and with what resources" (Fussel 2007, 268).

The second mode suggests that climate change assessment should be incorporated into city planning although some scholars have complained of insufficient data and techniques, as well as unsystematic incorporation. As suggested at the 2010 Cities at Risk (CaR) II conference, "there are current information/knowledge gaps and future research opportunities for addressing climate change related risks and vulnerability in Bangkok, HCMC, Jakarta, Manila and Mumbai" (Snidvongs 2010, 17). Furthermore, this approach should address "the need of institutional linking mechanism, assess the role of civic society groups and address deficiencies in existing planning instruments, in incorporating climate change risk and vulnerability" (18).

This problem also has emerged in developed countries such as Sweden. The use of climate knowledge in urban planning has already appeared in several projects, but there is a need to increase planners' knowledge and skills. There are still many barriers to doing so, such as the uncertainty and lack of arguments, communication problems between climatologists and planners, fear of complaints by other stakeholders, and the expense of climatic investigation (Eliasson 2000). The literature on this scenario of climate change,[1] however, still needs long, intensive processes to inform "the totality of knowledge about city's vulnerabilities and strengths" (Prasad et al. 2009, 8). Furthermore, the "current scientific (climate) information to develop better methods and procedures for designing the urban form" (Blakely 2007) should be factored into the planning process.

The third mode argues that adaptation planning depends on who or what is being adapted (Nelson et al. 2007). Thus, adaptation planning is applied on different scales. At the city level, using the examples of Durban and Quito, the cities' actions "were driven by internal goals and aims, rather than being pressured by mandates or agendas of external parties" (Carmin et al. 2012, 10). At the community level, "some societies may strive to adapt to climate change while maintaining a current standard of living–whereas others may aim to adapt simultaneously with improving the standard of living of their citizens" (Adger et al. 2009, 341). At the individual level, it could be someone who needs to retain the vitality and viability required to cope with the shock and/or adapt to the stresses caused by climate change. Therefore, the scales of space and time in adaptation planning differ at regional and individual levels, and they depend on who and what is involved in the adaptation process.

For both individuals and communities, adaptation planning does not have to depend on scientific climate proofing because little climatic data is available at the micro scale. Moreover, it is difficult to conduct any meaningful assessment of

---

[1] The Special Report on Emission Scenario (SRES) includes four types of storylines and scenario families (IPCC 2000). In Indonesia, SRESA1B is seen as the most suitable scenario because it implies that countries will balance their technological sources between fossil-intensive and non-fossil energy sources (Heru Santoso, Interview, 20 August 2012). However, the global circulation model (GCM), which represents physical processes in the atmosphere, ocean, cryosphere, and land surface, is the most advanced tool currently available for simulating the response of the global climate system to increasing greenhouse gas concentrations. To date, the GCM has succeeded in explaining the climate change phenomenon on regional and national scales, but it is limited in its application at the city and community levels.

changing climate conditions because climate modeling is still inaccurate (Firman et al. 2011). Therefore, the experiences of local people in changing climates are important in planning.

The Vietnamese who planted mangroves to prevent storm surges in the Mekong Delta (UNDP 2008) shared their insights with adaptation planners in Vietnam. The Papuan community that used their traditional tenure management system, known as *sasi*, conserved local biodiversity (McLean et al. 2009). The residents of Toineke Village in West Timor, Indonesia, have "adaptation strategies embedded in traditional practices and local knowledge" used those strategies to maintain their livelihood (Hornidge and Schotes 2012, 9). Hence, sharing information about planned adaptation practices should be socially constructed in order to disseminate the lessons learned. Laukkonen et al. (2009) recommended a methodology and an instrument to help individuals and communities in the planning process not only to gain their participation but also to embed their knowledge and actions into adaptation planning.

Attention to adaptation practices at the community level began several years ago. In 2012, IPCC's special report of the Intergovernmental Panel on Climate Change (SREX) noted the consensus and robust evidence for the inclusion of local knowledge in preparing adaptation policy. The community's knowledge and experiences should be integrated into the technical knowledge of climate change adaptation and disaster management. Some international donors have supported such initiatives, especially in poor communities (see UNDP 2012; CARE 2010; Mercy Corps 2011). They have developed several participatory models to conduct assessments and planning, most of which were intended to reduce vulnerability.

As a planning practitioner, I argue that community-based adaptation (CBA) is a new issue in the planning literature. In 2004, the UNDP reported that decentralized disaster risk planning in Haiti could be implemented feasibly and sustainably because the community-based mechanism already existed (UNDP 2004). Since then, the bottom-up approach to building systemic resilience to climate change through community-based adaptation programs has become imperative (UNFCCC 2009). The UNDP (2012) argued that CBA should be driven by community priorities, respond to location-specific needs, and offer lessons learned for the modified replication of best practices. In addition, the UNU-EHS suggested that "planning for meeting the requirements for the specific vulnerability groups and addressing contextual vulnerability at the local scale, rather than following top-down scenario based impact models, are also essential" (UNU-EHS 2011, 59).

It is not surprising that the subsequent initiatives of many community-based organizations (CBO) and nongovernmental organizations (NGO), supported by international donors and development banks, have organized projects on community-based adaptation planning. ADB (2011) supported a field test that integrated local knowledge and engaged four vulnerable communities in the Cook Islands to formulate adaptation plans. The Mercy Corps Indonesia (2010) used a Vulnerability and Capacity Assessment (VCA) of disaster preparedness and then incorporated it into a subdistrict plan. In both cases, the role of planners in community-based adaptation planning mediated the discussion between community groups and climate experts and connected local to scientific–technical knowledge.

The UNDP (2008) introduced Community-based Resilience Assessment (CoBRA), which integrates climate change risk management into MDG-focused initiatives. CARE (2010) also developed a toolkit for organizing community-based adaptation planning projects that result in climate-resilient livelihoods. The Pacific Institute developed an eight-step community-based research process that emphasizes the importance of recruiting and training participants (Garzón et al. 2012). The IISD (2012) developed the Community-Based Risk Screening for Adaptation and Livelihood (CRisTAL), a user manual and computer software program that produces planning scenarios based on climate-sensitive and/or natural resource-dependent livelihoods. Those initiatives plan community-based adaptation (CBA) based on scientific climatic information that is confirmed by local or traditional knowledge. The facilitator is the liaison between the experts and the community. Some toolkits for professional planners have also been developed. The Canadian Institute of Planners (CIP) (2011) published the Adaptation Planning Handbook. USAID (2009) published a guidebook for development planners on adapting to coastal climate change.

Most of these studies concentrate on rural areas, and they range from micro to macro levels. Such agencies use the sustainable livelihood framework and take a participatory approach to organizing planning projects. Most projects are designed to reduce vulnerability and improve relations between the community and the local government. However, to achieve a better understanding of the discourse of community-based adaptation planning, the scholarly literature requires other approaches and foci, especially urban areas where the vulnerable poor are concentrated.

Inspired by the notion of vulnerable place, most previous studies have used a place-based approach to investigating the adaptation to climate-related disasters. Many planning scholars, such as Hewitt (1997), Bahrainy (1998), Geis (2000), Birkmann (2005), and Blakely (2007), developed spatial planning models of reducing disaster risks. Hewitt (1997) linked land use, urban form, and urban design to the effects of natural disasters. Bahrainy (1998) developed a model of urban planning and design in a seismic region. Geis (2000) proposed an urban planning and design model by adding the notion of safe community but focused on to location of the community. Birkmann (2005) and Blakely (2007) addressed the dimension of the risks and resiliency of the community in spatial planning.

However, the discourse on the people-based approach still lacks empirical evidence of its contribution to making cities resilient. As stated by Geis (2000), "the only way to reduce the growing human and property losses from [s]evere flooding is rooted … in how we design and build our communities in the first place in these hazard prone areas" (153). The main issue concerns the kind of planning to be done. In 2009, the Commission on Climate Change and Development (CCCD) recommended the principles of human dimension in climate change adaptation for and beyond COP 15 Copenhagen. The CCCD (2009) argued that adaptation should consider human-based perspectives and apply them to the interfaces at local, national, and global levels in order to assess risks effectively. This change in perspective emphasizes the need for local adaptation strategies for engagement with learning gained in past experiences, the development of adaptive capacity by the people, the promotion of local capacities, the development diverse solutions, the promotion of the importance of ecosystem services, and the provision of public funding support for the poorest in society (Christoplos et al. 2009). Therefore, there is a need for adaptation planning that is driven by the local people who are directly affected by climate change.

2.3 The Nexus of Vulnerability and Adaptation: A People-Based Approach

The term *vulnerability* originates in the Latin word *vulnerabilis*, which means "being wounded." In the last decade, in the context of climate change adaptation and disaster risk reduction, concepts of vulnerability have evolved in terms of epistemological orientation, analytical framework, and methodological practices. Several social scientists have contributed to the shifting orientation of vulnerability, from exposure-oriented sources that require a physical system of protection to affected social entities that emphasize adaptation. According to Birkmann et al. (2010), early disaster risk reduction in the 1970s was still dominated by hazard perceptions, but in the 1980s, it was challenged by the vulnerability paradigm, which views vulnerability as the susceptibility and ability of both people and communities (Bankoff 2004) and the intrinsic predisposition to be susceptible to damage (Cardona 2004). Therefore, vulnerability is viewed not only as a given condition caused by disasters or climate change, but also as "the propensity of exposed elements to suffer adverse effects when impacted by hazard events" (IPCC 2012, 69).

In relation to climate change adaptation, the concept of vulnerability is based on how climate change affects communities and societies in different geographical regions (Bogardi et al. 2005). According to the Intergovernmental Panel on Climate Change (IPCC), vulnerability includes three elements: the sensitivity of a system to climate change, its adaptive capacity in relation to such changes, and its exposure to climatic hazards (IPCC 2007). Several scholars added the social dimension to define vulnerability in climate change adaptation. For example, vulnerability is "the predisposition of society and individuals towards a stressor or hazard to be harmed" (Wisner et al. 2004, 11). Furthermore, "the human system as an object of interest is vulnerable due to its own properties and stressors from nature, but also due to stressors from the human system itself" (Cutter 1996, cited in Fekete 2010, 18). Thus, according to Bengtsson et al. (2007), each social group has its own vulnerabilities, which "depend on people's interpretation through their personal life histories and experiences of daily interactions with the local environment" (Kuruppu and Liverman 2011, 658). Bankoff (2004) added that the subjective and intersubjective interpretation and perception of disaster events could determine the construction of the vulnerability of the city's inhabitants. In the *2013 World Social Science Report:*

2.3 The Nexus of Vulnerability and Adaptation: A People-Based Approach

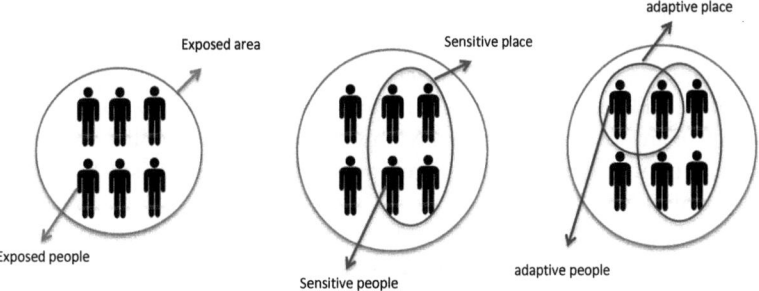

(i) Exposure and exposed people (ii) Sensitivity and sensitive people (iii) Adaptive capacity and adaptive people

Fig. 2.2 Vulnerability and vulnerable people (Source: Author)

Changing Global Environment, local knowledge was suggested for inclusion in policy-making:

> It is...important to go beyond the seeing is believing attitude typical of current evidence-based approaches to policy making. The accounts of the people who face environmental problems directly should also be accepted as valid. (Rajao) (ISSC 2013, 21)

Based on these concepts, we can understand that the level of vulnerability depends on how people perceive the impact on themselves and their properties. What they perceive is certainly specific and related to the kind of changes that they experience. The role of the affected people then is very important in understanding the meaning of vulnerability. Therefore, vulnerability not only positions human beings as the objects of vulnerability but also concerns who is vulnerable and how vulnerability is defined. The simple understanding that suggests that people who live in a vulnerable area are identical in their vulnerability is not relevant. The vulnerable area is derived from the conceptions of space and place, which delineate a certain location that is assessed to be sensitive, exposed, not adaptive, or maladapted. Moreover, vulnerable people are defined by the people-centered approach. Thus, it is imperative to differentiate vulnerable areas and vulnerable people (Fig. 2.2).

Two epistemological orientations form a spectrum of the complexity of vulnerability. At one end of the spectrum is the assumption that vulnerability can be reduced to its determinant factors through a causal relationship. At the other end of the spectrum is the belief that vulnerability should be viewed in a holistic manner. Fekete (2010) argued that reductionists analyze vulnerability using only one dimension; meanwhile, the holistic version synthesizes multivariables of vulnerability. Both views are very useful in identifying the object of interest. IPCC (2012) suggested that there are at least three objects: human beings, livelihoods, and assets. The characteristics of each object represent different conditions that are affected by hazard events. Therefore, a specific framework is required to define and measure vulnerability.

According to Birkmann (2005), vulnerability studies have evolved through the following: the sustainable livelihood framework (Chambers and Conway 1992), the double structure of vulnerability (Bohle 2001), natural hazard and risk (Bollin et al. 2003), the nexus of economic and social spheres (Turner et al. 2003), the pressure and release model (Wisner et al. 2004), the sustainable development agenda (UNISDR 2004), and the holistic approach to risk and vulnerability (Birkmann and Fernando 2008; Cardona 2004). The many approaches of vulnerability studies indicate that an integrated and cross-disciplinary perspective could be used to overcome the complexity of vulnerability.

According to Damm (2010), in methodological practices, vulnerability assessment is categorized into three groups: place-based vulnerability, which argues that vulnerability is determined by social and biophysical dimensions; multidimensional vulnerability, which argues that vulnerability is shaped by at least four dimensions—physical, environmental, social, and economic factors; and social–ecological vulnerability, which uses the interaction of social and ecological subsystems in defining vulnerability. Those concepts are derived from the holistic point of view of human–nature relationships.

The integration of the notion of social dimensions into the discourse of vulnerability also has emerged in climate change adaptation and disaster risk communities. Many scholars in various disciplines have discussed the social construction of vulnerability in order to arrive at a common and better understanding of who the vulnerable are and the kinds of adaptation options that should be provided. One determinant of vulnerability that has been discussed recently is adaptive capacity. Both climate change adaptation and disaster risk communities take into account causal relationship between vulnerability and adaptive capacity.

The present definition of vulnerable people takes into account the socio-spatial approach and begins by defining a vulnerable area. This approach considers social drivers, such as social norms, capital, or networks, in defining a vulnerable place. It identifies the households or groups that are located in the vulnerable place as vulnerable, in which age, wealth, gender, and social inequalities are recognized as differential factors (Christmann et al. 2012). However, socio-spatial approaches take into account the perceptions of the affected local people because they know their capacity to adapt to the stress and shock resulting from extreme climate events by using values and actions to increase or even decrease the sensitivity of their places and based their practices on their experiences of the exposure to climate-related disasters. Their everyday lives represent "a reality as interpreted by people and subjectively meaningful to them as a coherent world" (Berger and Luckmann 1967a, b, 19).

Adaptive capacity "plays a pivotal role in the progressive emergence of the vulnerability paradigm within the scientific realm" (Gaillard 2010, 223). The IPCC special report of 2012 indicates two ways of looking at the relationship: (1) vulnerability is the result of a lack of adaptive capacity, among other things, and (2) vulnerability is the opposite of adaptive capacity. The debate depends on which discipline and domain is used to examine the relationship. From the perspective ecological anthropology, which focuses on how cultural norms and practices reinforce human adaptation to the environment and how people utilize their cultural belongings to manage their environment, this relationship can be investigated in the realm of the local people who have lived experiences of environment changes, both gradual and rapid. I argue that the state of vulnerability can be viewed from the perspective of the population groups that experience adverse outcomes (exposed people), some of whom have the ability to adjust to environmental change (adaptive people). Therefore, the relationship between vulnerability and adaptive capacity should be explored from the point of view of individuals, not external experts.

Because it focuses on revealing individual experiences, the analysis of this relationship is better employed at the community level where the documentation of local people whose experiences and potential wisdom to decrease their state of vulnerability is easily found. Maguire and Cartwright's conception of the community includes three dimensions, which are considered in this research: a territorially bounded, socially connected, and/or shared interest group of people (Kumar 2005; Kelly and Adger 2000; Maguire and Cartwright 2008). Given the uncertainty of environmental change, communities that understand and respond to environmental changes will contribute to an inclusive and institutional mechanism of adaptation. However, the challenges in applying the findings need comprehensive generalization and detailed interpretation (Adger and Kelly 1999).

There are two views of the relationship between vulnerability and adaptive capacity. Some scholars have argued that adaptive capacity is a determinant of vulnerability. They positioned capacity under vulnerability, which is interpreted as a negative contributor (Wisner et al. 2004). Other scholars assume that capacity means that vulnerable people are not powerless (Bohle 2001; Gaillard 2010). Second, some scholars have argued that vulnerability is dynamic and determined by numerous factors. Therefore, it cannot be understood that low vulnerability is automatically reflected in high capacity (Alwang et al. 2001). This group tends to view capacity as the opposite of vulnerability.

According to this point of view, the relationship is socially constructed and leads to a further discussion of who is defined as vulnerable. Is everyone living in an affected area considered vulnerable? How do the affected people differentiate the vulnerable from the adapters? What kind of capacity is important for them? Those questions need to be answered by those affected by climate events. The examination of local people who experience the effects of climate change could elucidate this relationship, especially in communities in hazardous areas. These people certainly know how to survive under frequent occurrences of extreme climate events and disasters because they have lived there for a long time. This local knowledge needs to be revealed because it is essential to understand whether their ability to survive has transcended the state of vulnerability, which they define themselves. This clarification could resolve the shifting focus of adaptation policy by either reducing vulnerability or increasing adaptive capacity.

Theoretically, in order to define people who are vulnerable to floods, three conceptions derived from the vulnerability framework should be applied. The first is exposure, which means individuals or groups that could be threatened by floods. The second is sensitivity, which are the situations or conditions that render individuals or groups unprotected against floods. The third is adaptive capacity, which comprises the potential abilities of people or groups, which can be used to adapt to the changes caused by floods. At the conceptual level, the exposed population consists of individuals or groups who live in floodplains, low coastal areas, riversides, and land subsidence areas that are inundated. These people tend to be poor, disabled, and elderly. Adaptive people have the technical skills, experience, and resources to adapt to floods.

Consequently, if we take into account the perceptions of these individuals and groups, we may conclude that unlike invulnerable people, vulnerable people are exposed, sensitive, and not adaptive. From the conceptions, we can assume that the capacity of people is crucial in determining whether someone is vulnerable. The centrality of capacity is the key factor in managing the negative changes that come from both the outside (exposure) and inside (sensitivity). If people do not want to be exposed, they should be able to move or to build defenses. If people want immunity (are not sensitive), they ought to have the capacity to strengthen themselves through frequent practices or other measures. Hence, adaptive capacity can prevent or remove vulnerability if people overcome their exposure and sensitivity. Adaptive capacity is used not only to mitigate hazards and their magnitude in terms of disaster management or to protect people against the effects of climate change but also to address the multidimensions of vulnerability in the context of sustainable development.

Therefore, the points of view of vulnerable people are an important starting point in adaptation planning, where the focus is not only on reducing vulnerable areas but also on increasing the adaptive capacity of vulnerable people. We need to shift the focus of vulnerability assessment in order to define both vulnerable regions and vulnerable people. In global discourses, many vulnerability studies are based on socio-spatial concepts rather than on human capacity. The state of vulnerability needs to be assessed based on the experiences of the people who are exposed, sensitive, and not adaptive to environmental change. Other interests should not influence the label of vulnerability (Table 2.2).

The definition of vulnerable people can be traced to discussions of the vulnerability framework (see Table 2.3), which began in the 1990s. Chambers and Conway (1992) used the sustainable livelihood framework to analyze social sustainability in a rural area, especially how farmers coped with shocks and stress. These two scholars focused on livelihood assets that were affected by shock or seasonality changes, and they found that governmental intervention was important in increasing social sustainability. Subsequently, Bohle (2001) examined the role of the "social assets" of vulnerable households or groups in coping with risks and shocks. Wisner et al. (2004), with Pressure and Release (PAR), developed a three-level model of vulnerability that included root causes, dynamic pressures, and unsafe conditions. They argued that people would become vulnerable if they lacked access to power, structures, and resources and were marginalized politically and economically. In the second level, vulnerability occurs when there is a lack of local institutions and when macro forces pressure people to adapt. In the third level of this model, people have unsafe physical and social relationships, a weak local economy, and unstable public institutions. Using the disaster risk index (DRI), the UNDP (2004) tried to evaluate vulnerable people using the ratio of the number of the exposed population to the frequency of

2.3 The Nexus of Vulnerability and Adaptation: A People-Based Approach

Table 2.2 Identification of vulnerable people in vulnerability frameworks

Framework	Main factors of vulnerability	Main factors of adaptive capacity	Pertaining of vulnerable people	Authors/supporters
Sustainable livelihood framework	Livelihood assets, shocks, trends, and seasonality	Government system; power relation	(rural) community and society	Chambers and Conway (1992), DFID (2005), Haan and Zoomers (2005)
Duo-systems	Internal factors (coping capacity) and external factors (risks and shocks)	Government policy and community responses	Households and groups	Bohle (2001), Cannon et al. (2003), van Dillen (2004)
Pressures and Release (PAR)	Progressive interrelation among root causes, dynamic pressures, and unsafe condition	Politic and economic system	National and local entities	Wisner et al. (2004)
Human and environmental linkages	Exposures: Coping, impact, and adaptation responses	Institutions, political economy, transitions	Global, national, and local institutions	Turner et al. (2003)
Disaster risk management	Physical, social, economic, and environmental factors	Physical planning, social and economic capacity, and management	General society	Davidson (1997), Bollin et al. (2003), UN/ISDR (2004)
	Mortality and economic losses and damages	National and global cooperation	Countries	UNDP (2004), World Bank (2004)
Holistic approach (BBC framework)	Physical exposure, socio-economy fragility, and lack of resilience and abilities	Institutional structures and public policies and actions	Not directly mentioned; probably from national to local entities	Birkmann and Fernando (2008)

Source: Author, based on Birkmann et al. (2010)

physical exposure. The World Bank (2004) released a "hotspot" index that generalized the number of people at risk based on mortality, economic loss, and GDP proportion as determinants of vulnerability (Bogardi et al. 2005). In conclusion, these authors described vulnerable people as a group, but not as individuals. They attached multidimensional conditions to vulnerable people but not the endogenous capacity of the individuals.

These authors tended to use socio-spatial global and local entities to assess the role of adaptive capacity in reducing vulnerability. Turner et al. (2003) stressed the

Table 2.3 Approaches to vulnerability

Vulnerability Elements	Eco-place-based approach	Socio-spatial-based approach	People-centered approach
Exposure	Floodplain area, river banks, etc.	Fishermen village, farmers village	Poor, women, children, etc.
Sensitivity	Ecosystem services, coastal area, etc.	Slum area, dense settlement	Individuals, households and groups
Adaptive capacity	Urban system, rural system, etc.	Social institutions, local markets, power relation, etc.	Experiences, local innovation, creativity, self-reliance

Source: Author

roles of institutions, political economy, and the transition from global to local entities. According to Bogardi, Birkman, and Cardona's (BBC) framework, vulnerability reduction is a dynamic condition of social, economic, and environmental dimensions in different times and institutions within a feedback loop system. Therefore, being proactive before a disaster is as important as is being capable in managing emergency responses and recovery phases (during and post disasters).

However, the focus of those assessments is not on the people who have lived through the disaster events or have adapted to climate change. The original picture of vulnerability should be derived from the people who have lived through disaster or adjusted to environmental changes. The first enabling source of adaptive capacity is the locally situated knowledge gained from that experience. Moreover, adaptation should depart from previous definitions of what people need to improve their lives. Thus, the paradigm of people centeredness includes vulnerability and adaptation and examines the nexus of vulnerability and adaptation.

Furthermore, Birkmann (2005) suggested that vulnerability studies should be placed in the context of development to improve adaptation. In this research, I use Korten's view, which focuses on creativity and initiative, to define adaptive capacity. I assume that people who are exposed and sensitive to disaster would become vulnerable if they lacked adaptive capacity. In contrast, people who optimize their capacity would reduce their vulnerability. Therefore, I argue that vulnerable people should be viewed as not only victims but also sources of creativity for developing their own adaptive capacity. They are the actors "who define the goals, control the resources, and direct the processes affecting his or her life" (Gran 1983, 176). Thus, I aim to develop a phenomenology of vulnerability in the paradigm of people-centered development. This paradigm will bring an alternative approach to the assessment of vulnerability and the provision of adaptation options (See Table 2.4). It shifts the focus of vulnerability assessment from the environment to the lived experiences of humans.

People should build their own adaptation pathways to transform passive masses to proactive citizens. The plan occurs on a small scale, perhaps using a technology suited to their culture. Through the assistance of an alert citizenry, self-organizing systems could be developed through people-scale organizational units and self-

Table 2.4 The difference between place-based and people-centered approaches

Substance	Place-based	People-centered
Goal	Spatial quality	Human livability
Drivers	Visionary	People needs
Process	Interdisciplinary	Transdisciplinary
Scale	Spatial scale (remote sensing)	Personal scale (human sensing)
Experts main	Geography and engineers	Anthropologists and sociologist
Knowledge	Technical	Experiential
Main issue	Spatial dynamics	Human right

Source: Author

reliant communities in adapting themselves and their area to environmental changes. As argued by Bob Doppelt, who is in the resource innovation group, "if we focus on experiences, people will be in a setting where they can grasp the idea of interdependency and reciprocity" (Garrison Institute 2013, 14).

2.4 The Framework of People-Centered Adaptation Planning

As I explained in the previous chapter, I apply the people-centered approach, which is used in development studies, to examine the adaptation pathway of the urban poor in Jakarta. As defined by David Korten (1990), development is "a process by which the members of society increase their personal and institutional capacities to mobilize and manage resources to produce sustainable and justly distributed improvements in their quality of life, consistent with their own aspirations" (Korten 1990, 67). Therefore, the people-centered approach "looks to the creative initiatives of people as the primary development resource and to their material and spiritual well-being as the end that the development process serves" (Korten and Garner 1984, 201).

In the planning discipline, the importance of centering people in the planning process has been recognized since the early 1990s. According to John Friedmann, humanist philosophies evolve in planning practices. He argued that the planning discipline has adopted philosophical and social values such as "advocacy of the poor and other marginalized people, citizen participation, inclusiveness, and the right to housing" (Friedmann 2000, 7). At the operational level, Friedmann (1984) developed the concept of a territorial unit to define inclusive development strategies based on the principle of self-reliance. In his case study, he defined an individual as the member of a territorial unit that could extend from neighborhood to a nation. The territorial unit is based on residential place rather than temporary economic utility. In line with Friedmann, Korten argued, "the well-being of all its members in perpetuity is a main concern when delineating the territorial jurisdiction" (Korten and Garner 1984, 210). Therefore, by adopting the territorial unit in the planning discipline, people can express and share the knowledge learned from their lived experiences.

The wide spectrum of the planning field, which ranges from technical design to public policy, has led to several discourses on how planning can be useful to and owned by people. In his *Cities for People*, Jan Gehl, a popular Danish architect, recalled the importance of the scale of people in planning the city (Gehl 2011). He suggested that planners should shift to a plan-by-people perspective for the good of people. In the sociopolitical context, planning has already brought people into the planning process because they have rights and knowledge. John Forester's *Deliberative Practitioner* and Susan Fainstein's *The Just City* are examples of progressive planners' acknowledgment of the role of community deliberation to make planning socially accepted. Therefore, I argue that making people central to the planning process brings human values into the planning practices.

The location of people in the anthropogenic landscape varies from one region to another. As part of Southeast Asia, Indonesia is an archipelagic country of various cultural and traditional values. In the context of urban settlement, the presence of *kampungs* is a unique characteristic of Indonesia's urban development. Although the definition of the *kampung* as a slum[2] is still debated, most Indonesian scholars emphasize that the *kampung* is a traditional form of urban settlement (KBBI 1991) where residents live in the same way as in their original villages (Wiryomartono 1995). According to Mielke (2007, 13), who has discussed the multitude of local conceptions in defining the village in Kunduz province, Afghanistan, "the local concepts of village are quite different from the western idea and that they area contested." Localities such as the *kampung* are present in other parts of Asia (Marcussen 1990), such as the *bustee* in India, a type of officially authorized slum (Kundu 2003) and the *katchi abadis* in Pakistan, "a sub-standard settlement that generally occupied government land with no legal property rights" (Fernandes 1994, 51). Most of those scholars argued that the absence of the governmental management of such localities has produced these informal settlements. Therefore, it is imperative to consider the specific locality of communities during the planning process.

As a self-constructed settlement, the *kampung* represents the Indonesian spirit of *gotong royong* (mutual self-help and exchange), which differentiates it from other types of urban settlement. Jellinek (1999, 267) argued, "mutual self-help whose physical fabric has evolved organically–creating a sense of place." In the everyday life of the *kampung*, *kerja bakti* can be translated as the embodiment of *gotong royong*. It includes "traditional institutions" (Wilhelm 2011, 123), "collective activity" (Lont 2005, 42), and "duty work" (Perkasa and Hendytio 2003, 130).

---

[2] *Kampung* is associated with slums and squatters because the dwellers are low-income people. According to Suparlan (2004), slums are identified by improperness, irregularity, communality, social economic stratification, and informality. Squatters are the denizens of slums that are illegally built in prohibited locations and public spaces. The recognition of the *kampung* as a slum appeared through the Kampung Improvement Program (KIP) in early 2000, which was supported by the World Bank and UNCHS (World Bank 1999). Both agencies denoted the *kampung* as a slum because it had the same typologies as slums or squatter settlements, which are irregular, self-made residences on land unsuited for residence, such as floodplains, swamps, riverbanks, toll roads, and railway areas.

2.4 The Framework of People-Centered Adaptation Planning

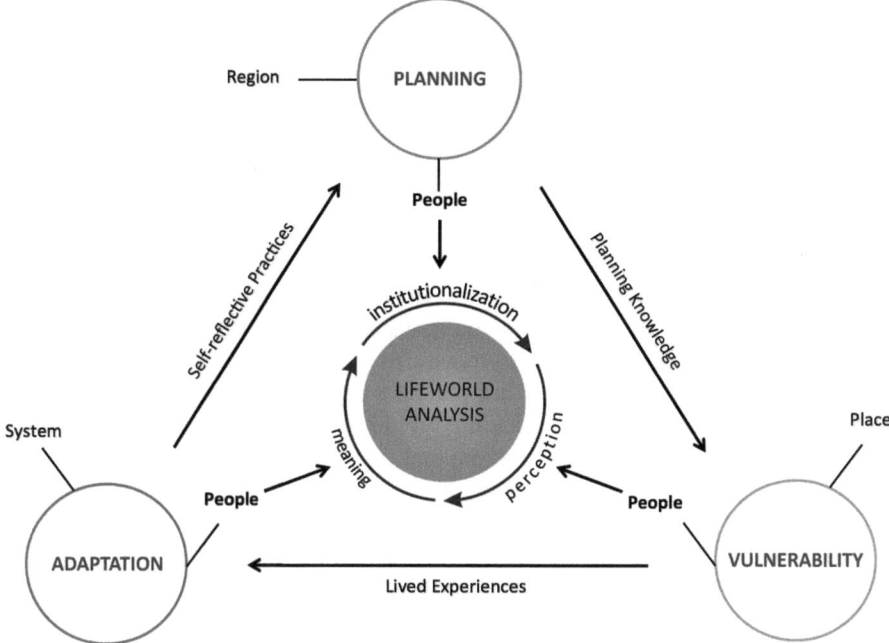

Fig. 2.3 People-centered adaptation planning framework
This figure is a trilogy that shows the linkage between vulnerable, adaptation, and planning. The vulnerable people use lived experiences to adapt and therefore to be adapters. The adapters apply self-reflective practices to plan their adaptation pathways. They institutionalize the knowledge of planned adaptation within their community. This knowledge will help vulnerable people to adapt to future floods. This trilogy provides a robust thesis on how adaptation planning can proceed and be institutionalized at the local level without the label of "climate proofing", but based on individuals' practices.

I argue that the *kampung* should not be perceived merely as an urban problem but as way to make cities resilient. The residents of the *kampung* are the subjects of the kind of adaptation pathway that they have built today and will build in the future. How they define vulnerability and identify vulnerable people and adapters, how they practice adaptation and reflect those practices, and how they institutionalize the latter in the community will be discussed in the following sections. Therefore, the *kampung* should be viewed as a research laboratory that produces deeper knowledge about urbanization (Hornidge and Antweiler 2012).

In this book, I develop a framework that links the conceptions of vulnerability, adaptation, and planning (see Fig. 2.3) from a people-centered perspective. I analyze the social construction of reality. I use Schütz's *lifeworld* theory to reveal the lived experiences of *kampung* residents and examine the relationships among those concepts. By using *lifeworld* to analyze the trilogy of vulnerability, adaptation, and planning, this research will determine the adaptation pathway of the *kampung*.

I conceptualize the flood-related vulnerability based on the everyday lives of the urban poor who live in the *kampung* to identify the vulnerable and the adapters.

Next, I explore the practices of the adapters in the context of adaptation to regular floods, which is rooted in climate change adaptation and disaster risk reduction. I want to know the meaning of what they have practiced to date and how they prepare for future floods. I examine the processes using reflective practice theory. Lastly, I disclose the way they institutionalize their adaptation to develop a planning model that is appropriate to the localities that participate in the planning process at the community level.

Before I elaborate these concepts, I reiterate that this research is rooted in the concept of phenomenology in planning, and it can be placed in the debate of people-centered and place-based planning. This debate is ongoing because although the approaches are different, they complement each other. This debate emerges from the endeavor of planning theorists to transform their ability to understand the real world. Friedmann (2008) argued that planners have the cognitive limitations of scale, complexity, and time. After reading Friedman's essay and reflecting upon my 12 years of professional practice, I agree that the planning discourse will continue to evolve in its understanding the changing world in order to create a better place in which to live.

There are two kinds of planning (see Table 2.4). Place-based planning perceives urban planning as territorially bounded because the urban space is multi-scale. The scale of urban space can be mapped. The planning map is the representation of reality from time to time and provides a simplistic model of urban complexity. It involves technical contributions from other experts, such as geographers, economists, engineers, and even sociologists. Therefore, the academic and professional learning process is imperative in producing the knowledge of planning.

The people-centered approach perceives that planning is rooted in the realm of the people who live in the city. Beginning with Jane Jacobs', *The Death and Life of The Great American Cities* (1960), several planners have tried to make their paradigm people centered. Susan Fainstein's *The Just City* (2000) and Jan Gehl's *Cities for People* (2011) demanded that the paradigm of planning shift to accommodate "a human-centered philosophy" (Friedmann 2008). Fainstein (2005) suggested, "planning theory needs to consider under what conditions conscious human activity can produce a better city (region/nation/world) for all its citizens" (127).

As suggested by John Friedmann, planning theory is "a work of translation that raises the horizon (of planning) to include the vast field of human knowledge or perhaps, of the knowledge in the plural" (Friedmann 2008, 29). This research is a voyage taken to discover the concepts and ideas in the sea of Schütz's *lifeworld*, which I translate into the language of planning when I return to my work as an urban planner. I search for planning knowledge that is embedded in the realm of the urban poor who experience regular floods and who are neglected in the discourse of adaptation planning. The practical knowledge of situational planning requires quick and practiced adaptation, and it can be learned by rote. According to Scott (1992, 315), it is "a metîs, a contour of practical knowledge that can be acquired only by local practice and experience." A metîs of planning is subjective, contextual, particular, and often consists of tacit knowledge, therefore differing from planning science, which is impersonal, universal, and concerned with certification process (Scott 1992). Therefore, the "knowledgeability of individuals" (Antweiler and Mersmann 1996, 13) plays a key role in this research.

Chapter 3
Planning Institutions of Adaptation to Flood in Jakarta

Abstract I discuss the contextual features of adaptation planning in Jakarta. I explain the planning institutions that are constituted in Jakarta regarding flood management, and how government regulations and the initiatives of other stakeholders work to take responsibility for the effects of floods. I found that although different planning approaches have been taken. The different realms of planning would produce inefficient and ineffective adaptation and potentially create other new problems. Furthermore, I argue that urban adaptation planning in Jakarta should also focus on community empowerment and rely on the local knowledge that is embedded in each community.

Keywords Jakarta • Adaptation • Institution • Planning

I discuss the contextual features of adaptation planning in Jakarta. I explain the planning institutions that are constituted in Jakarta regarding flood management, and how government regulations and the initiatives of other stakeholders work to take responsibility for the effects of floods. I found that although different planning approaches have been taken. The different realms of planning would produce inefficient and ineffective adaptation and potentially create other new problems. Furthermore, I argue that urban adaptation planning in Jakarta should also focus on community empowerment and rely on the local knowledge that is embedded in each community.

3.1 Understanding Floods in North Coastal Jakarta

More than half of the population of Indonesia, which is one of the fastest urbanizing countries in Southeast Asia, now lives in urban areas. DJPR (Bahasa Acronym for Directorate General of Spatial Planning) predicted that eight of ten Indonesians will live in cities by 2050 (DJPR 2013). Fifty percent of all cities in Indonesia, or about 47 cities, are coastal. Sixty-two percent are secondary cities, and only five cities, including Jakarta, are metropolitan. In coastal cities, floods are serious problems, in addition to coastal erosion and seawater intrusion (DJPR 2013). According to the

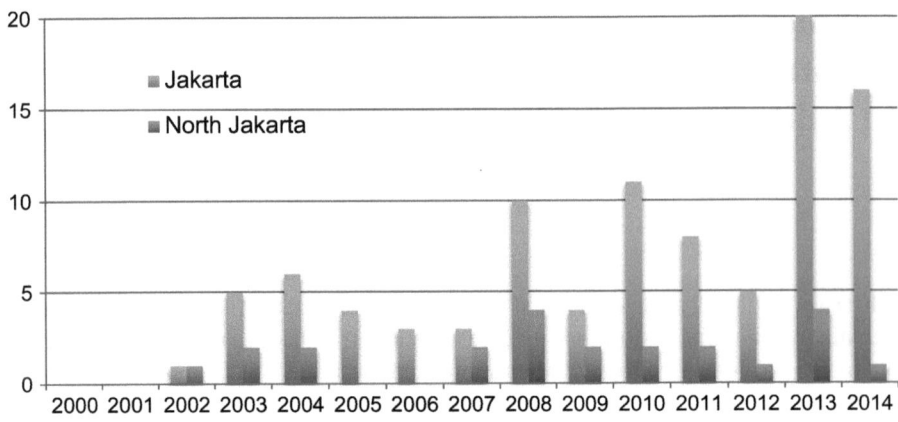

Fig. 3.1 Number of floods in Jakarta (Source: the author, based on BNPB (2014))

Table 3.1 Rainfall during two major floods of Jakarta

Rainfall data	2001–2002	2006–2007
Total rainfall for five stations (mm)	7100,0	7484,0
Maximum rainfall upstream (mm/day)	168.1	247,0
Maximum rainfall downstream (mm/day)	172.0	234.7
Average rainfall for five stations (mm/dy)	21.1	25.8
Average rainfall upstream (mm/day)	20.6	24.8
Average rainfall downstream (mm/day)	21.9	27.3
Percentage of days with rainfall	69.9	67.0
Duration of event (days)	121.0	88,0
Water level at Manggarai (cm)	1050,0	1061,0
Flood level at Bukit Duri (m)	2.3	3.4

Source: World Bank (2010)

National Agency of Disaster Management (BNPB), in 2014, there were 131 floods in Jakarta and 22 floods in North Jakarta in the rainy season (see Fig. 3.1).

Since the seventeenth century, when it was Batavia, Jakarta has faced floods (Soehoed 2004). *Kompas* daily news on 14 September 1998 reported large floods in 1671, 1699, 1711, 1714, and 1854 (as cited in Julianery 2007). According to Soehoed (2004), the flood was caused not only by the overflow of river runoff but also by heavy precipitation. Precipitation is still one of the main causes of floods. The data on major floods in 2002 and 2007 gathered by MGK (*Bahasa* acronym for the Meteorology Office of Indonesia) shows that the total volume of rainfall affects the water level in Manggarai and flood level at Bukit Duri (see Table 3.1). Rainfall intensity, not the number of rainy days, tends to be a cause of major floods.

In period from 1980 to 2012, the data showed a fluctuating pattern of rainfall (Fig. 3.2). Although the slope line tends to be stable, there was much more rainfall

3.1 Understanding Floods in North Coastal Jakarta

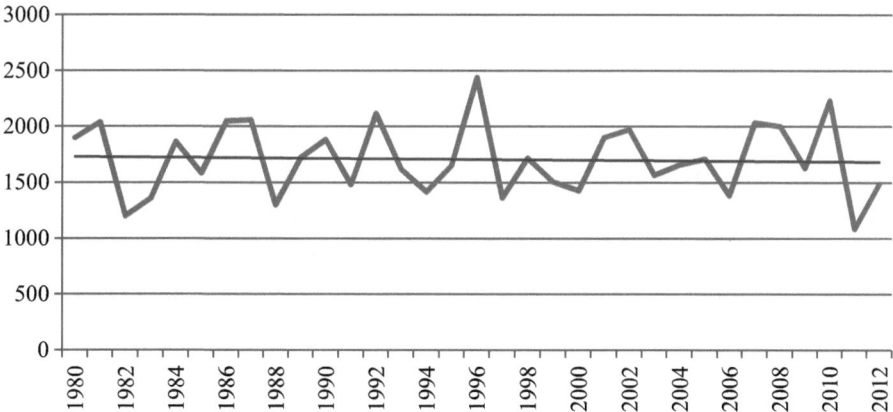

Fig. 3.2 Observed rainfall in the period 1980–2012 (Source: the author, modeled by Sutikno)

in 1996 (2400 mm), 2007 (2000 mm), and 2010 (2200 mm). In these years, several large floods occurred in Jakarta. Therefore, the climate change in Jakarta is not simply a matter of more rain but the uncertainty of heavy rains. Instead, the mean of the rainfall tends to decrease slightly. Thus, the trend in the rainfall pattern is unrelated to the level of floods, but heavy rainfall is the major cause. The intensification of rainfall and alterations in seasonal cycles also significantly influence the risk of flooding. Climate change is predicted to result in an increase in annual rainfall by 2–3% in Indonesia (Ratag 2001). Climate change also affects the duration and occurrence of La Niña and El Niño, which later influence the rainfall and other patterns. During La Niña, Indonesia has more rainy days in a year; during El Niño, there is drought or a shorter rainy season (fewer rainy days in a year).

In addition to heavy rains, the slow onset phenomenon that contributes to floods is sea level rise (SLR), which is influenced by the combination of melting of glaciers and polar ice caps and the thermal expansion of ocean water that inundates low-lying coastal areas (PEACE 2007; Muhammad 2011). In 1990, the ITB research team found that the SLR in Jakarta was 4.38 cm per year. In 2007, based on data collected in the period from 1984 to 2006, the SLR was 7.00 cm per year (Aldrian 2007). In 2010, Plamonia calculated that the mean SLR increase in Jakarta Bay was 5.75 centimeters (cm) per year. SLR is found in coastal Jakarta. The widening gap between the static surface of the sea and land has increased the velocity of seawater infiltration, which is exacerbated by heavy rainfall.

Another sea phenomenon that contributes to flooding is the anomalous tidal levels caused by the earth's gravity and the ocean's circulation. In Jakarta, most people call these *banjir rob* or tidal floods. They usually occur in the rainy season and in the seasonal transition period, particularly during the full and new moons (Hildaliyani 2011). In the last 7 years, there has been no clear pattern of these periods or the frequency of tidal floods. The biggest repetition occurred in 2009 when there were seven tidal floods, but there have been at least three per year since

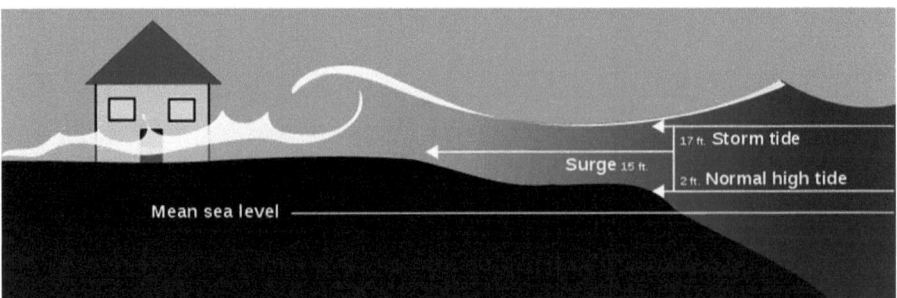

Fig. 3.3 Illustration of tidal floods (Source: Brinkman 2012)

then. The only similar pattern is repeated location, such as Muara Baru, Tanjung Priok, and Penjaringan, which experience periodic tidal floods. There have been no occurrences of tidal floods in the same month or on the same days in each year (Appendix 5).

The Jakarta Coastal Defense Strategy (JCDS) argued that tidal floods happen "when the Java Sea rises during the monthly lunar tidal cycle" (Prameshwari 2009), which is an 18.6-year cycle (astronomic tide). The peak of the next tidal cycle is expected in 2025 (Brinkman 2012). The JCDS predicts that the area will be at the mercy of an upswing in the tides unless a sea defense system is built by 2020. In the 2007 flood, huge amounts of water blasted inland and inundated most of the North Jakarta area. Moreover, the tidal flooding is worse if the rain-induced flood is accompanied by high spring tides. The flood could last for 3 days (*Kompas* 2013) (Fig. 3.3).

The low elevation of coastal North Jakarta also contributes to tidal flooding. Lying under the sea surface ranging from −1 to −3 m (Abidin et al. 2008), the coastal area is vulnerable to high spring tides (Brinkman 2012) and rising sea levels. By 2050, the average inundation will range from 0.28 to 4.17 cm. There are 24 inundated areas, which comprise about 30% of the coastal area, while half this area is complicated by water scarcity caused by seawater infiltration (Meliana 2005).

The occurrence of natural factors, heavy rains, SLR, and tidal floods confirms that the coastal flooding of Jakarta is caused by climate change. Lobo (2008) found that the contribution of high precipitation and rising sea levels was significant in increasing the proportion of flooded areas in Jakarta. He predicted that the inundated area of Jakarta would increase 50.21% by 2035; half of Jakarta will be inundated. Many scholars have argued that the North Jakarta municipality is vulnerable to climate change, such as "high tides coupled with growing rising sea water" (Firman et al. 2011; Susandi 2009), rainfall (Yusuf and Fransisco 2009), and rising sea levels in addition to land subsidence (Ward et al. 2010). Therefore, the *Dewan Nasional Perubahan Iklim* (DNPI) claims that Indonesian coastal cities will continue to experience the effects of climate change in at least these four major ways: increased sea surface temperature and rising sea level, increased frequency, intensity

3.1 Understanding Floods in North Coastal Jakarta

of extreme weather, and changes in seasonal cycle and rainfall pattern (cited in Muhammad 2011).

In addition to these climate factors, manmade factors have raised flood levels. The main factor is the land subsidence caused by groundwater. According to Ward et al. (2010, 7), "there is a strong indication that land subsidence in Jakarta is highly correspondent with the high volume of groundwater extraction from the middle and lower aquifers." It has been reported that the rate of groundwater extraction rapidly increased by 40% per year from 1998 to 2007 (*Kompas* 27 September 2010). Between 1982 and 1997 (15 years), land subsidence reached 0.2 m although the first notice was recorded in leveling surveys conducted in 1926 and 1927 (Abidin et al. 2008).

In another study, Colbran (2009) found that from 1993 to 2005, the land in the north coastal area was reduced by −0.57 m. Priyambodo (2005) predicted that the rate of land subsidence per year was about 0.8 cm. The excessive use of groundwater and the volume of groundwater extraction have already exceeded the land's capability to recover (Samsuhadi 2010). It will affect the lower land surface of Jakarta, mainly on the coastal side. Gumilar et al. (2009) argued that the land subsidence could cause tidal flooding, tilt houses and other buildings, and, at worst, cause gas pipelines to explode.

According to Abidin et al. (2008), the load of construction (i.e., settlement of high compressibility soil) and the natural consolidation of alluvium soil are the main factors that trigger the groundwater extraction. The extensive physical development caused by uncontrolled rapid urbanization has stressed Jakarta's carrying capacity, including its groundwater resources. Furthermore, uncontrolled urbanization has disrupted land use patterns. Many floodplain areas have been occupied by both formal and informal settlements. The expansion of the informal settlements often intrudes upon riverbanks, lakeside, and green open spaces. Other water catchment areas channel an overflowing river to another area. In addition, real estate companies have reclaimed land from the sea in areas such as *Pantai Indah Kapuk (PIK)* and *Taman Impian Jaya Ancol*. It is obvious then that floods in coastal Jakarta are exacerbated not only by climate change but also by human activity.

Regular coastal flooding has adversely affected the economic and social life of Jakarta residents. The flood risk constantly increases because the flood-prone areas continuously expand and interfere with the economic activities of Jakarta. Hallagate et al. (2013) estimated that by 2050, the annual average losses (AAL) of Jakarta would have increased by 54% from 2005 or about 0.22% of gross domestic product (GDP), even with stringent adaptation planning. This figure is higher than predicted for Bangkok (0.09% of GDP) and lower than that predicted for Ho Chi Minh City (0.83 of GDP). An estimated 98,877 people were evacuated in 2002, and 17,000 were evacuated in 2007 (Rujak Center 2011). Two thousand were evacuated in the most recent flood in 2013 (Revianur 2013). These events brought urban activities to a halt for days. The floods have caused water scarcity, disease outbreaks (Haryanto 2009), and dozens of deaths. Furthermore, most victims of these events are among the urban poor.

According to the government's statistical data, in North Jakarta, which has approximately 1,645,659 inhabitants or about 17.13% of total population of Jakarta,

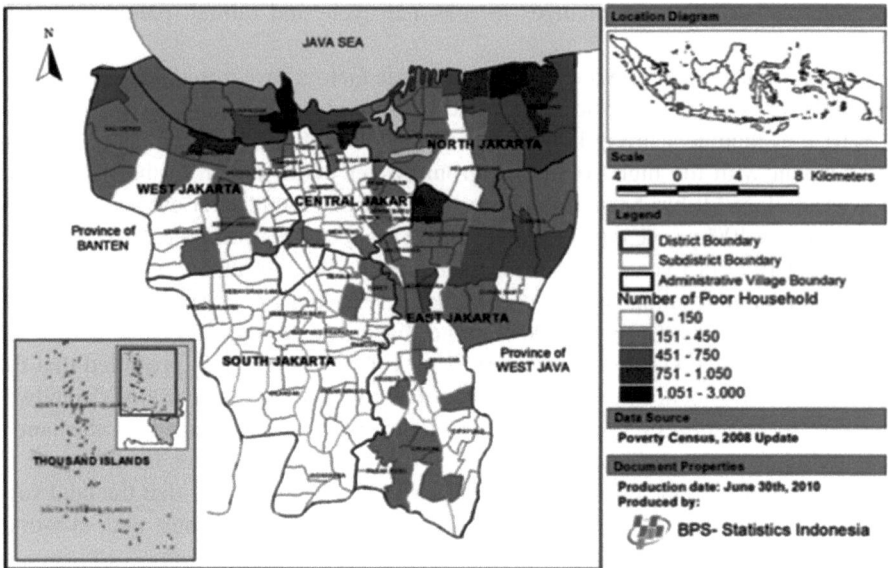

Fig. 3.4 The distribution of poor households in Jakarta (Source: Firman et al. 2011)

27% are poor (BPS 2010; DKI Jakarta 2010). Figure 3.4 shows the distribution of poor households in Jakarta. Based on the poverty data released by the Central Bureau of Statistics (BPS) in 2008, North Jakarta and East Jakarta have the most concentrated areas of the poor. In North Jakarta, *kecamatan*[1] Penjaringan is the most vulnerable to floods induced by rising sea levels because it has the highest concentration area of poor and nearly poor, who are vulnerable to becoming poorer if there is a disaster (Firman et al. 2011). In 2011, the World Bank noted that the homes and livelihoods of the poor are the most vulnerable to floods.

3.2 City-Level Adaptation Planning: Infrastructure-Driven Approach

The planning policies and programs for managing floods in Jakarta have changed since the period of Dutch colonialism ended in 1945. These policies and programs have been linked to institutional changes in Jakarta. The discourse of flood management planning can be divided into five periods: pre-independence, *Orde Lama* (1945–1965), *Orde Baru* part one (1965–1985), *Orde Baru* part two, and reformation (since 2000). Each period has specific problem diagnostics, regulations, planning knowledge, leadership, and funding resources. Consequently, the ever-changing planning institutions have slowed the implementation of planning for floods (see Table 3.2).

---

[1] Kecamatan is an administrative subdivision of a city (kota) or a district (*kabupaten*).

3.2 City-Level Adaptation Planning: Infrastructure-Driven Approach

Table 3.2 History of flood management planning in Jakarta

	1900s–1945	1945–1965	1965–1985	1985–2000	2000–2013	2013–now
The main flood plans	Infrastructure drainage rescue and health planning	Infrastructure planning + tourism development (promenade)	Inter-sectoral planning (flood to water supply and poor settlement)	Integrated infrastructure planning, attached to the spatial planning	Comprehensive, integrated to spatial planning	Master plan of NCICD—National capital integrated coastal development
Regulation	Van Breen plan 1922 (transversal channel)	Pluit polder plan 1957	Masterplan for drainage and food control 1973	Masterplan of flood infrastructures 1997	Perda 1/2013 regarding the spatial plan of DKI Jakarta 2013–2023	Ongoing, probable presidential regulation
Leaders	General governor of Batavia	Mayor of great Jakarta (Sudiro)	Governor of DKI Jakarta Province (Ali Sadikin)	Suprapto (governor of DKI Jakarta Province)	Sutiyoso and Fauzi Bowo (governor of DKI Jakarta)	Boediono (vice president RI) and Joko Widodo (governor)
Resources/partners	Dutch colony	City budget supported by national budget	Supported by loan OECF/JIBC + French protocol loan	Supported by JICA and World Bank	Supported by World Bank and JICA	Dutch government
Planning area	Old Batavia (central and west part)	North Jakarta	City of Jakarta	City of Jakarta + Puncak	Greater Jakarta	Coastal and sea area
Causal factors	Overflow rivers and high precipitation	Increased rainfall was not stored by drainage system	External rainfall	Complex; multifactors	Complex; inter-regional problem	Complex, exacerbated by climate change and land subsidence
Big flood events	1671, 1699, 1711, 1714, 1854, 1921	1956	1970, 1976	1991, 1993, 1994, 1996	2002, 2007	2013

Source: Authors, based on Julianery (2007), Lobo (2008), Fitrinitia (2012), Yusuf (2013)

The shifting diagnostics of the causes of floods have made planning interdisciplinary. In the Dutch colonial era, the flood infrastructure was designed only to reduce the effects of the floods on public health. The plan now capitalizes on the flood infrastructure for economic reasons, such as using a seawall as a pedestrian walkway, including floodwaters in the water supply, and adding recreational functions to the retention pond. Planning knowledge has also moved from physical drainage systems to a comprehensive plan that incorporates land and water management. It has also affected the adaptation planning for flood events.

The only similarity between the planning of the past and the present is infrastructure. For many decades, flood management in Jakarta has been dominated by the flood infrastructure. The leading government agency is still the Department of Public Works. However, the scale of planning has grown from the municipal to the regional, and the institution responsible is now the National Ministry of Public Works.

For more than 100 years, the national and city governments have been implementing flood management plans, but flooding remains unpredictable. Although planning processes have involved many international experts, and they have been supported by numerous donors and international agencies, floods remain a chronic problem in Jakarta. Several experts have stated that Jakarta will be destroyed unless there is stronger control over land development. The destruction of Jakarta has been under public discussion since 2010 (see Table 3.3).

The long experience of flooding has shaped the perceptions of Jakarta's urban stakeholders. As a capital city where national treasures are located, as a megacity that is the headquarters of multinational companies, and as a giant city with more than ten million residents of all income levels, Jakarta has become a witness to and an object of numerous flood initiatives. All stakeholders have their own realms of interest according to which they decide what adaptation options they support.

According to the Intergovernmental Panel on Climate Change (IPCC), "adaptation is the adjustment in ecological, social or economic systems in response to actual or expected climatic stimuli and their effect" (IPCC 2001). Adaptation consists of the development of technology or infrastructural changes that maintain livelihoods and the actual behavioral adjustments to adapt livelihoods to new climatic conditions. Some solutions may only be short term and are limited by their lack of flexibility, and some others have long-term goals. Therefore, adaptation must be planned.

How macro-adaptation is planned depends on the needs and the capacity of the urban stakeholders. For the government of DKI Jakarta Province, an adaptation plan is needed to outline the annual programs and to estimate the budget. The government agencies are then able to identify funding sources and provide mechanisms to defray the adaptation costs. However, in making the plan, the government excludes non-state actors who may have interests in or have already conducted an adaptation. For example, some new town developers have protected their environment from floods by building their own dykes, water pumps, and drainage systems. These developers have even recruited international experts to formulate an adaptation plan, and they have invested their capital in building the required flood infrastruc-

3.2 City-Level Adaptation Planning: Infrastructure-Driven Approach

Table 3.3 News regarding flood Jakarta

No	News title	Newspaper	Time	Groups of actors
1.	*Krontjong Toegoe, Tafsir Jakarta Tenggelam*	Kompas	20.01.2014	Experts (archeologist)
2.	*Jakarta diramalkan tenggelam karena punya patahan aktif*	Viva News	18.01.2014	National government
2.	*Hampir separuh wilayah Jakarta bakal tenggelam di 2050*	Kompas	26.12.2013	Journalist
3.	*27 Januari, Jakarta "Tenggelam"?*	Kompas	20.01.2013	Scientist (hydrology)
4.	*Jakarta "tenggelam" sudah di depan mata*	Kompas	27.09.2010	Scientist (geodetic engineer)
5.	*Jakarta Tenggelam*	Kompas	18.09.2010	Public figure (former minister)
6.	*Menteri PU: Jakarta terancam tenggelam*	Kompas	31.07.2010	National government
7.	*WALHI: Tahun 2050 Jakarta tenggelam*	Kompas	13.02.2009	NGO

Source: Author, based on newspaper clipping

tures. However, urban stakeholders, especially urban poor who lack the capacity to adapt their settlements and livelihood to flood, had not been taken into account (JICA-RI 2011). In many cases, they have already conducted adaptation planning because they have experienced floods, and have incorporated their experiences into the planning of macro-adaptation.

The government regards a flood as a preventable disaster. Numerous studies of flood impacts and risk assessments have been conducted to improve the rigorousness of urban development planning. Therefore, both the national and DKI Jakarta provincial governments have prioritized strategies for adaptation to floods through the development of a system that protects Jakarta against floods. At least four planning documents focus on the adaptation to floods in the coastal area: *Rencana Tata Ruang Wilayah* (RTRW) DKI Jakarta Province (spatial planning), *Perencanaan Kembali Kawasan Pantura Jakarta* (re-planning study for new development of Jakarta coastal area), *Rencana Aksi Daerah-Adaptasi Perubahan Iklim* (RAD-API) DKI Jakarta Province (action planning for climate change adaptation), and *Rencana Penanggulangan Bencana* (RPB) of DKI Jakarta Province (disaster management planning).

RTRW 2010–2030 supports the adaptation to floods in coastal Jakarta by integrating the flood infrastructure into the water resources infrastructure. RTRW classifies the infrastructure as water conservation, water utilization, and water-damaged power control. RTRW also defines the coastal area as a provincial strategic development area with regard to environmental concerns. Article 77 states that coastal development is driven by sea dykes and island reclamation (DKI Jakarta 2011). Thus, further detailed spatial planning is required.

The national government has been planning Jakarta's coastal development since 1995. However, because of the 1997 economic crisis and the Ministry of

Table 3.4 History of coastal reclamation planning

1990–1994 (preparatory program)	1995–1998 (establishment program)	1999–2003 (delaying program)	2004–now (replanning and implementation)
Held seminars and conferences both national and international	Governor presentation to the president of Republic of Indonesia	Vacuum due to economic crisis of 1997	Environmental impact assessment
Assisted by Dutch and Australian technical consultant	Coordination with the Ministry of Home Affairs and related government agencies, regarding the institutional arrangement	The regulation of governor of DKI Jakarta Province No. 138/2000 regarding guidance of reclamation	Economic feasibility assessment
	Presidential Decree No. 52/1995 regarding land reclamation	Decree of Ministry of Environment No. 14/2003 regarding unfeasibility of coastal reclamation of Jakarta	The Decree of Supreme Court Reg. No. 12PK/TUN/2011 regarding regranted the reclamation of Jakarta
	Regulation of government of DKI Jakarta Province No. 8/1995 regarding spatial planning		Principle permit of reclamation for central and west zones
	The regulation of governor of DKI Jakarta Province No. 973/1995 to 220/1998 regarding acting agency of coastal Jakarta development		Regulation adjustment to new law regarding spatial planning and decentralization
	Feasibility study and partnership cooperation with private developers		

Source: the author

Environment's rejection of reclamation planning in 2003, the reclamation program has been delayed (see Table 3.4). The planning of coastal Jakarta was resumed in 2010. The national government has implemented new laws and regulations, ensured the commitment of private developers, taken into account global and local environmental challenges, and promoted a new vision of Jakarta as a service city. This replanning of north coastal Jakarta has integrated sea reclamation with the revitalization of the mainland and the conservation of the *Kepulauan Seribu* district.

In 2008, a national seminar on north coastal Jakarta was held at the University of Indonesia. The governor of DKI Jakarta claimed that the replanning of coastal Jakarta had a threefold goal. At the global level, the goal was to increase Jakarta's competitiveness as a service city and place it on par with other global cities. At the regional level, the replanning aimed to reduce the urbanization of the upstream area (Bogor/Depok), build a new growth center for megacities such as Jakarta, and overcome the bottleneck of a watershed system. At the local level, the replanning aimed to improve the environmental quality and social welfare of North Jakarta, reduce the traffic in the city center, and develop a new image of Jakarta. The vision of the new Jakarta through the replanning of the north coastal area implied that the government had transcended the problem and offered new economic opportunities to Jakarta's people.

Rencana Penanggulangan Bencana (RPB) is a disaster management tool that is required at the national, provincial, and city levels. It reflects a dual shift in the focus of disaster management from recovery to prevention and from government to governance (Simarmata and Suryandaru 2015). It is based on the Presidential Regulation No. 21/2008 and is part of development planning (Article 6). Based on this regulation, the government of DKI Jakarta Province was facilitated by *Badan Nasional Penanggulangan Bencana* (BNPB) or the National Agency of Disaster Management to produce the *Rencana Penanggulangan Bencana* (RPB) in 2011. Subsequently, the government of DKI Jakarta Province arranges RPB for its cities and districts. However, at present, the RPB of North Jakarta City is still being conducted. In relation to floods, the RPB of DKI Jakarta Province consists of two objectives: (1) the distribution of flood risk areas in the Jakarta region and (2) management planning to solve the problem.

The *Rencana Aksi Daerah-Adaptasi Perubahan Iklim* (RAD-API) DKI in Jakarta province enables stakeholders to integrate and synergize adaptation in short and long terms in order to achieve resilience (de Boer 2013). RAD-API consists of five sectors: energy, public health, infrastructure and settlement, ecosystem services, and slum and coastal areas. In relation to floods, the RAD-API DKI predicts that because of rising sea levels, the areas inundated by floods will increase (see Fig. 3.5). Therefore, RAD-API recommends the same flood infrastructure plan that was proposed by *Rencana Tata Ruang Wilayah* (RTRW). In the health sector, RAD-API

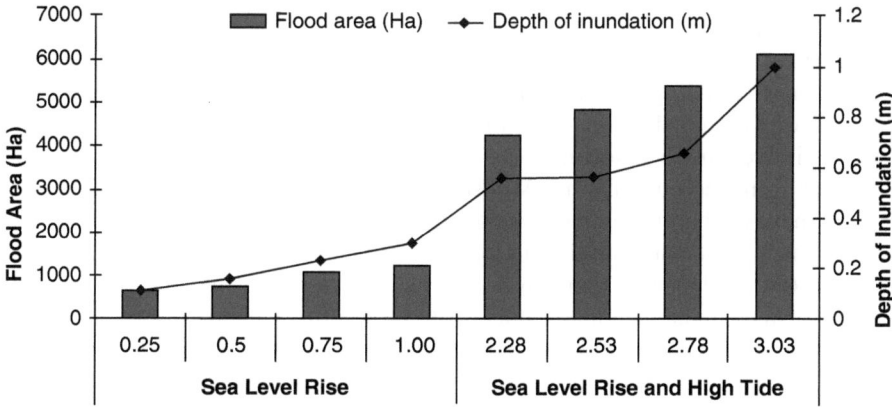

Fig. 3.5 Flood areas due to rising sea level (Source: de Boer 2013)

also plans to reduce potential vector-borne diseases that could be triggered by climate change. In the ecosystem services, RAD-API plans to rehabilitate catchment areas and improve natural ponds and rivers.

In addition to this flood mitigation plan, RAD-API describes several plans related to institutional arrangement that need to support the development of flood infrastructure, such as strengthening flood control regulations and promoting the research and development of science and technology for climate adaptation. Moreover, RAD-API included the need to increase the knowledge capacity of slum dwellers and coastal communities in managing climate risks.

Although the four plans have different recommendations for adaptation, the planning goals have the same long-time perspectives and common interests in protecting the urban space (Table 3.5). Based on the planning documents, the government has not concentrated on the population that is vulnerable to floods. The government assumes that by reducing or eliminating the flood hazard, people can live safely. Therefore, they prioritize physical engineering solutions to reduce inundation and to predict future floods.

The planning institution for flood management was incorporated into the Presidential Regulation Number 54 Year 2008 regarding the spatial plan of Jabodetabekpunjur, which includes Jakarta, Bogor, Tangerang, Bekasi, Puncak, and Cianjur. This planning umbrella promotes and accelerates intercity cooperation, especially neighboring cities such as Depok, Bogor, and Tangerang. Therefore, the planning for flood management has shifted from inter-sectoral cooperation to multilevel governmental collaboration.

The macro perspective on adaptation planning has been discussed in academia. Most scholars have argued that climate change has exacerbated Jakarta's flood risk (Firman et al. 2011; Fuchs 2010; Yusuf and Fransisco 2009). The *Institut Teknologi Bandung* (ITB) projected that the sea level on the waterfront of Jakarta will rise by 0.57 cm per year and that most of North Jakarta area will be submerged by 2050 (Firman et al. 2009). In line with this argument, many theses have supported the influence of rising sea levels on Jakarta's floods (Lobo 2008; Sentosa 2010; Rahmat 2011). Based on such research, scholars have suggested that adaptation to floods should be part of the adaptation to climate change.

Considering these facts, I concede that the Jakarta adaptation plan, especially with regard to flooding in North Jakarta, has captured national attention. In January 2013, Dewan Nasional Perubahan Iklim (DNPI), or the National Council of Climate Change, invited me to a meeting to discuss the coastal development of Jakarta. The presenter from the coordinating Ministry of Economy introduced the broadened development of the giant seawall by adding further island reclamation to accommodate the high demand for global properties and national infrastructure. As shown in Fig. 3.6, the adaptation planning for North Jakarta has integrated the new giant seawall with the reclamation land not only to protect this vulnerable area but also to implement new standards of Jakarta development. The strategy to capitalize on the risky area of North Jakarta gave the impression that the modernization of Jakarta would take place by simultaneously securing and developing the city.

3.2 City-Level Adaptation Planning: Infrastructure-Driven Approach

Table 3.5 Planning related to flood of government of DKI Jakarta Province

	RTRW DKI Jakarta	Replanning pantura Jakarta	RPB DKI Jakarta	RAD-API DKI Jakarta
Planning goal	Shaping urban structure and regulating land and water uses	Integrating coastal development program	Managing the disaster events, from pre- to post disaster stages	Reducing the vulnerability due to climate change
Planning horizon	20 years	20 years	20 years	20 years
Planning approach	Spatial analysis	Multi-sectoral integration	Four-disaster cycle management	Climate trends analysis
Relation to the adaptation to flood	Defining floodplain zone	Identifying coastal defense strategy	Mapping the flood risk area	Identifying climate change impact both slow onset and rapid changes
	Promoting or controlling the development rights of land and water bodies.	Regulating the coastal reclamation development	Identifying the prevention, emergency response, raising awareness, and recovery and reconstruction program	Offering adaptation options to those impacts
	Identifying flood mitigation infrastructure			
Statutory	*Peraturan Daerah* DKI Jakarta Province number 1/2013 (province regulation)	*Peraturan Gubernur* (governor regulation)	*Peraturan Gubernur* (governor regulation)	*Peraturan Gubernur* (governor regulation)
Lead organization	*Dinas Tata Ruang* (urban planning department)	*Bapppeda* (planning agency)	BPBD (disaster management agency)	BLHD (environmental management agency)

Source: the author

We therefore infer that the adaptation planning for floods in Jakarta has become less defensive and more aggressive. The government of DKI Jakarta Province has shifted the locus from the land-based defense of rivers, canals, water pumps, and lakes to sea- and land-based protection, including reclamation, pond retention upstream and downstream, and the seawall (see Fig. 3.6). However, the shift from defense to development has overlooked the presence of the vulnerable urban poor. The government seems to want to respond to the global challenges and intense competition by transforming the image of Jakarta's north coast. In other words, the government of DKI Jakarta prioritizes the adaptation of the city structure and the protection the residents from floods instead of improving the human capacity to

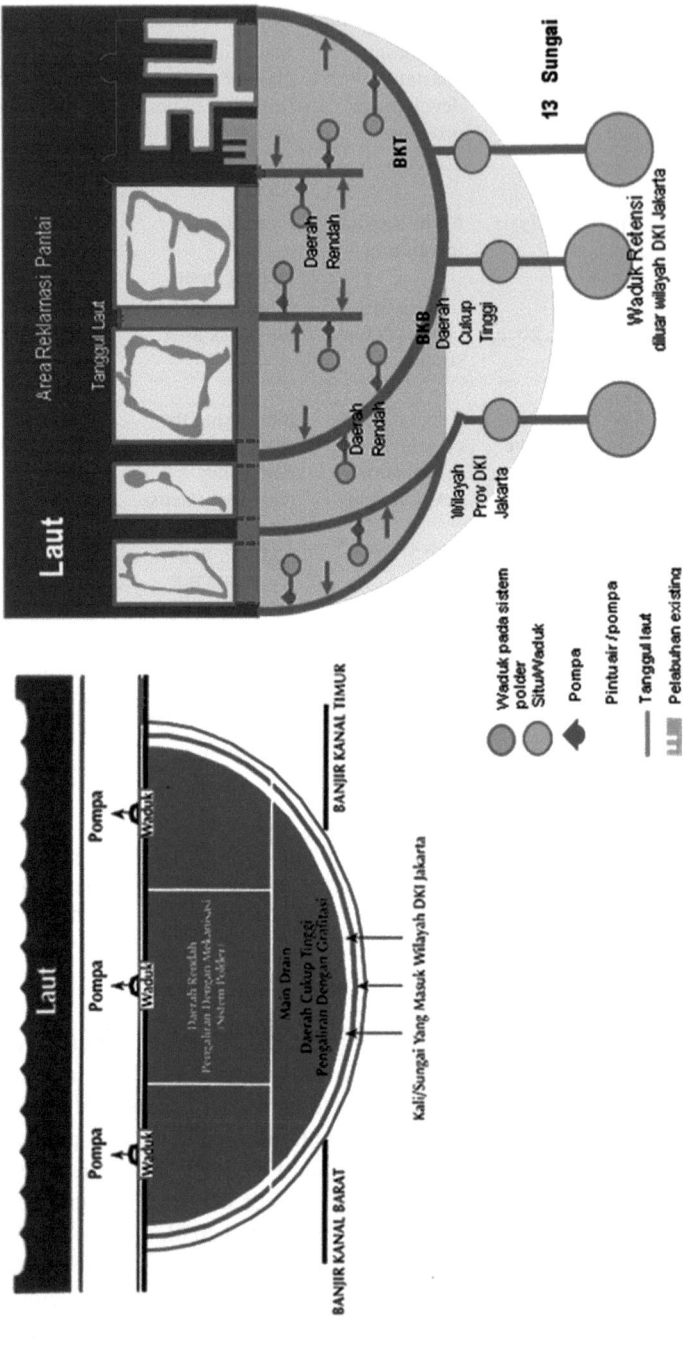

Fig. 3.6 Shifted planning concept of flood control (Source: Bowo 2010)

3.2 City-Level Adaptation Planning: Infrastructure-Driven Approach

adapt. Sea reclamation is perceived as a rational response to the rapid urbanization that burdens the upstream and downstream watershed areas.

Infrastructure-driven planning of flood management, which has progressed from the local to the regional scale, is apparent in the mid-term development plan of DKI Jakarta Province (RPJMD DKI Jakarta). In the period from 2007 to 2012, the plan promoted cooperation with the neighboring cities of Bogor, Depok, and Tangerang through the East Canal development, which connects the drainage systems of Bekasi and Jakarta and the normalization of the Ciliwung River, which crosses Bogor-Depok-Jakarta. In the present period from 2013 to 2018, RPJMD DKI Jakarta has included the development of polders, retention ponds, a greening program, and infiltration wells in Bogor, Cianjur, and Depok. The content of this mid-term development planning evidences that flood management planning for Jakarta has extended to neighboring cities.

The issues of climate change adaptation and disaster risk reduction have also influenced infrastructure-driven planning. The increasing flood risk caused by climate change, such as rising sea levels, high tides, and heavy rains, has transformed the land use planning and zoning regulations of the coastal area of Jakarta. The integration of island reclamation and seawall developments has been proposed as a solution to flood management. At the beginning of 2014, the government of Indonesia initiated the Master Plan of National Capital Integrated Coastal Development (NCICD) and established the Project Management Unit to develop an integral solution to the problem of flood management. The NCICD future image (see Fig. 3.6) tries to shift the mindset from avoiding the risk to manage the risks and even creates a new development. This living dyke concept has also created issues of environmental justice along the coastal line of Jakarta, which need to be considered by Jakarta when developing the urban adaptation plan.

The possible destruction of Jakarta by flooding has motivated the government to plan a flood control system. The government spends much energy on flood infrastructure development on the mainland, such as a canal system, river dredging, water ponds, and drainage improvement, as well as the development of sea walls. Even though several experts have found that the root of the flood problems is the public compliance with the land use planning in both downstream and upstream areas, the government has maintained its focus on the flood protection system in the Jakarta area. Moreover, climate change has provided an additional reason for improving and enhancing the flood control system.

> Jakarta will be thoroughly inundated by 2050, and the height of water could reach 50 cm [over the land surface] and it only comes from the global warming impacts, not considering storm surges and spring tide activities. Kecamatan Penjaringan, Pademangan, Cilincing, and Tanjung Priok will be the most vulnerable. (Hadi 2010)[2]

The government's reasons for prioritizing the physical infrastructure were confirmed in a high-level policy dialogue held by the Indonesian Association of Planners (IAP) in November 2011. The forum highlighted the importance of the program to

---

[2] Cited in *Republika Online*, 25 September 2010; translated by the author.

protect national vital objects, such as power plants, fuel logistics depots, and other public assets from the threat of rising sea levels and tidal floods. The presenter from the government of DKI Jakarta Province and the head of the Department of Public Works argued that the government had extended the height of the seawall from 0.8 to 2.5 m for a distance of 32 km (*Kompas* 2011) and prepared the development of a giant seawall which was scheduled for construction in 2015 to protect those public assets. In addition, the national government, which was represented by *Badan Perencanaan Pembangunan Nasional* (Bappenas) or the National Development Planning Agency, stated that the losses and damage caused by the 2007 floods reached IDR 5.2 trillion,[3] not including the economic losses of private companies and insurance companies that were estimated at IDR 3.6 trillion[4] during the 7–10 days of the flooding. Sarman Simanjorang, Vice President of the Chamber of Commerce and Industry of DKI Jakarta Province, stated that the daily losses caused by the 2014 flood in Jakarta were about IDR 100 billion[5] (Toyudho 2014).

The government of DKI Jakarta Province considers the expansion of urban slums and squatters on the riverside, lakeside, and coastal area a significant cause of floods. The informal settlers have significantly reduced the amount of land previously used for a green open space included a retention pond and absorbed rainfall water. The huge volume of garbage that the settlers generate has lowered the capacity of the rivers and the retention pond. This uncontrolled spatial development is a determining factor in Jakarta's floods (Brinkman 2012; Firman et al. 2011). Therefore, moving the slum dwellers from the area is part of the government's plans to make the urban structure resilient to floods.

As explained by the officer of Spatial Planning Department of DKI Jakarta Province, providing affordable housing for slum dwellers is currently the best option because land tenure prevents the consolidation of this area. However, the government of DKI Jakarta admitted in an interview, "relocating the urban slum dwellers is not an easy task." Land tenure and the housing regulation are major problems in managing the slum area, which was emphasized in an interview with Izhar Chaidir, Acting Secretary of Spatial Planning Department:

> The land problem… [It is a complex [problem]. One of the causal factors is the *pronas* (national program in the economic crisis era [1999], which prohibited people from occupying abandoned land). It had become an interstice for certain people to certify the land, later generating the problem of double certificates… [It was also problematic at the first time since] there was a precedent that the land had been illegally rented by several tenants and housekeepers and was becoming a business commodity. (Interview with Izhar Chaidir, Acting Secretary of Spatial Planning Department, May 2012, translated by the author)

> [In order to overcome this problem], we began to build cheap apartments to attract people to move. It worked to some extent. In the past, we had *rusunami* [cheap apartments for low-income dwellers], but it failed because they used them as a business commodity. No more *rusunami*, it is a rent-house now. [However], there is still a precedent that the housing units

---

[3] IDR 5.2 trillion = 346.7 million Euro (1 euro = IDR 15,000).
[4] IDR 3.6 trillion = 240 million Euro.
[5] IDR 100 billion = 6.7 million Euro.

have been illegally rented by several tenants and housekeepers. So, we realize that the target group should be rotated, from one slum area to another, interchangeably… We give the incentive fully, free one-year lease, fully furnished…. There is progress. More people are willing to live in cheap apartments. (Interview, Izhar Chaidir, Acting Secretary of Spatial Planning Department, May 2012, translated by the author)

When I met him again in August 2013, he reiterated that moving slum dwellers into affordable apartments was better than letting them struggle in their current location. When I asked whether there were opportunities for a *kampung*, with its original sea-based culture, to have a legal space in Jakarta Bay and be integrated into the new island reclamation program, he replied that it was not suited to the spatial law and was inconsistent with the new vision of Jakarta Bay. The newly elected governor of DKI Jakarta Province had launched the *kampung* readjustment program (*kampung deret*), but it only applies to *kampungs* that have a clear land status. Therefore, I concluded that the government has two plans for the adaptation pathway of the *kampung*: developing a flood infrastructure and then relocating the *kampung* to cheap apartments in safer places on land and the floodplain zone.

When I traced the references used by government(s) in prioritizing the adaptation plans, I found that the sources were donors and academics. The latter have conducted several studies[6] over the last 10 years. From the donors' perspective, Jakarta lacks the infrastructure to absorb, store, and channel water when heavy rainfalls are exacerbated by high tides. They blame Jakarta's floods on the poor synchronization of planning and implementing institutions (see Fig. 3.7).

Based on studies conducted from 2007 to 2009, the Jakarta Emergency Dredging Initiatives (JEDI) and the Jakarta Urgent Flood Mitigation Project (JUFMP) recommended the enlargement of two canals, the normalization of 13 rivers, and the improvement of the drainage system. From 2010 to 2014, through the Jakarta Coastal Defense Strategy (JCDS), the Dutch government emphasized that the flood hazards caused by the sea were exacerbated by the land subsidence caused by groundwater abstraction. The JCDS recommended three stages of planning, all of which focused on the development of a defensive infrastructure. In the first stage, sea and river dikes, retention ponds, and pumping stations would be reinforced. The second and third stages would be concerned with the combination of the infrastructure with the reclamation program and the utilization of the sea dike as transportation infrastructure. Hence, "defense" has been replaced by "development." This planning concept suggested that Jakarta should build a flood infrastructure to pro-

---

[6] According to PvW (2010), numerous donor studies have recommended the improvement of drainage system and spatial planning of Jakarta, such as the Quick Reconnaissance Study Flood JABODETABEK in 2002 by Rijkswaterstaat in 2002; Flood Management Study for DKI Jakarta Report by Rijkswaterstaat, in 2003; Drainage Management for Jakarta—Priority Assistance—DKI 8 by Louis Berger Inc. in 2004; Outline Plan for Major Drainage and Small Lakes Management in the Jabodetabek-Bopunjur Area, Pusat 3.10 by Nippon Koei and Kwarsa Hexagon (in 2005); Jakarta Flood Management: Non-structural Measures by the consortium of Witteveen and Bos, Royal Haskoning, WL/Delft Hydraulics, HKV, Euroconsult Mott MacDonald, and DHV in 2007; and Integrated Planning for Space and Water by the consortium of TU Delft, Demis BV, WL/Delft Hydraulics, and MLD in 2008.

Fig. 3.7 Illustration of the future vision of North Jakarta development (Source: Kemenkoeko (2014))

tect the city and reduce the vulnerability of North Jakarta, which covers a double sea dyke, canal, river dredging, water ponds, and a reclaimed island (see Elings 2011; Prasad et al. 2009).

Donors have been instrumental in shaping Jakarta's adaptation planning. The knowledge transmitted by these donors is based on the macro-view of floods, in which both donors and the government perceive that the adaptation pathway on Jakarta flood should depart from the hydrological problem. Such adaptation requires the improvement of the structure of watershed management based on an integrated engineering solution. This transmitted knowledge also appears in the shift of adaptation planning from a defensive strategy to a development strategy, which requires technological tools. The knowledge was transmitted from abroad because Indonesia still does not have the modern technology to protect against floods. Lastly, the history of development partnership in the flood management sector shows the dependency of the Indonesian government on international agencies. The government of DKI Jakarta and the national government had signed several memorandums of understanding (Droesch et al. 2008) or partnership agreements with donors, foreign governments, and international NGOs (i.e., World Bank, JICA, and the Dutch government).

Partnerships with donors and NGOs in adaptation planning are not new. A global survey conducted by MIT in 2012 showed that the adaptation planning in many cities consisted of forming not only advisory groups but also partnerships with NGOs, other cities, businesses, and community groups. Figure 3.8 shows that "Asian cities appear to frequently seek out partnerships, while they are generally uncommon in Australia and New Zealand" (Carmin et al. 2012, 18). Therefore, it is not surprising that Jakarta seeks the advice of donors and international organizations (Fig. 3.9).

3.2 City-Level Adaptation Planning: Infrastructure-Driven Approach

---

Concept note of Jakarta Emergency Dredging Initiative (JEDI) in 2007

List of the following institutional root causes of flooding in the JABOTABEK area:

- Lack of enforcement of regulations on groundwater abstraction.
- Lack of enforcement of spatial plans and building regulations.
- Limited coverage of solid waste collection services. Insufficient funding for operations and maintenance.
- Limited technical expertise.
- Lack of enforcement of forest law and regulations.
- Insufficient funding for investments in new flood control infrastructure.
- Lack of coordination between authorities responsible for water resources management.
- Lack of incentives for interregional coordination.
- Absence of political leadership to address the above issues in integrated manner.

Source: JCDS, 2010: 5

Fig. 3.8 The concept note of Jakarta Emergency Dredging Initiative (JEDI) in 2007

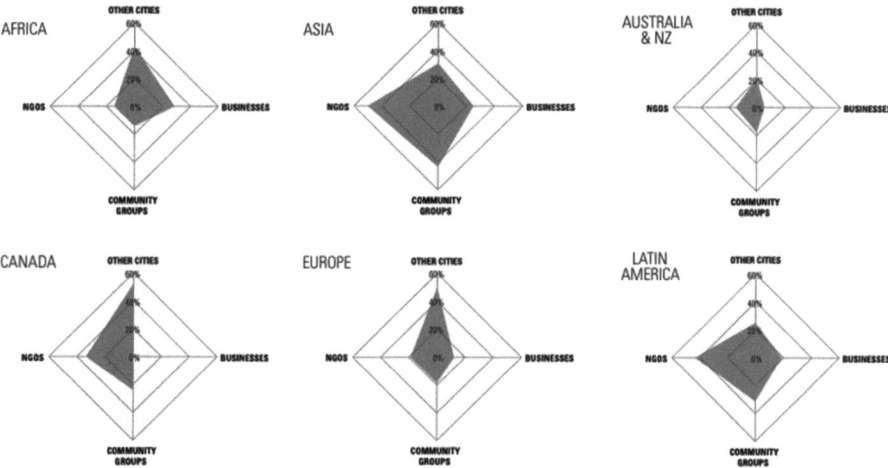

Fig. 3.9 Partnership by region (Source: Carmin et al. 2012)

The recent infrastructure-driven planning of Jakarta shows that government agencies still concentrate on preventing the flooding of Jakarta. Although governments have planned many flood infrastructures over centuries, floods still inundate Jakarta. This contradictory picture has certainly led to different perceptions of floods. The public discussion still centers on how Jakarta could live without floods. Nonetheless, the concentration on protecting Jakarta has influenced urban stakeholders' adaptation to floods. The donor's influence on the idea of infrastructure-driven solution has been identified for long time. Furthermore, flood issues have moved from urban environmental concerns to the political agenda. The government has shifted the infrastructure plans from protecting the coast to developing sea reclamation.

3.3 Unrecognized Knowledge of Adaptation Planning at Community Level

The problem of Jakarta's floods should also be seen from a micro perspective. Because of the differences in flood magnitude and frequency in several parts of Jakarta, floods have varying levels of significance. The coastal communities are the first to be flooded. They have their own adaptation plan, but they still need support from the government. The government must consider its adaptation pathway. The World Bank (2010) indicated that the government of DKI Jakarta has much to consider in addition to the construction of a flood infrastructure. It also has to plans for climate-related disasters within communities. The World Bank Indonesia recommended strengthening local institutions by understanding the adaptation actions taken by the urban poor and community because "there is a wealth of knowledge and many strong existing social networks within the vulnerable communities of Jakarta" (World Bank 2010). These could be accessed by empowering community-level administrators, such as *Lurah*, heads of RTs, and heads of RWs in organizing community-level activities and by learning what RTs and RWs have done and what they need in order to continue their program. However, some RTs and RWs are stronger than others are.

The Ministry of Oceans and Fisheries (MOF) supports the empowerment of villages. The MOF has endorsed *Desa Tangguh* or "resilient village" development in 6640 villages and in Jakarta, where two of 15 villages are supported. The program focuses on five dimensions: human resources, natural resources, livelihoods, infrastructure and environment, and awareness of disaster and climate change. However, NGOs influence the adaptation process at the community level. Therefore, participative planning has become the main tool used for the adaptation planning by NGOs.

Mercy Corps Indonesia (MCI) is an international NGO that is concerned with climate change adaptation (CCA) and disaster risk reduction (DRR) in many Indonesian cities. It has created a Local Resilience Action Plan (LRAP) "to stimulate actions at the local level and initiate a better dialogue between community, sub sub-district (*kelurahan*), provincial level government and other stakeholder to implement the plans" (Shah and Ranghieri 2012, 34). MCI uses a transdisciplinary approach to develop methods for making LRAPs, which consists of sensitization, technical analysis, stocktaking and assessment, option identification and program prioritization, and plan creation. Thus, LRAP is a climate change participatory planning document formulated by community and *kelurahan* (local government).

Even though the World Bank financially assists these tasks, the respective communities are expected to own the LRAP. In 2010, MCI developed LRAPs in three *kelurahans* in Jakarta: Pluit, Kapuk Muara, and Pademangan Barat. MCI formed a working group of *kelurahan* officials, informal leaders, youth groups, community organizations, and women's organizations. MCI aims for LRAPs to function as a reference for the implementation of climate change adaptation activities by local stakeholders and other parties in the respective *kelurahan*s. LRAP is also expected to stimulate actions at the local level and initiate a dialogue among communities, *kelurahan*, provincial level governments, and other stakeholders (Mercy Corps Indonesia 2010).

However, this goal may not easy to achieve. Pramita Harijadi, the task leader of Mercy Corps, conducted different vulnerability and capacity assessments (VCA) in order to determine the usefulness of LRAPs for the communities. The current LRAP is one of the pilot studies she organized in addition to the ACCCRN program in Lampung and Semarang. According to Pramita Harijadi, LRAP Jakarta needs to combine the results of VCAs, which were conducted by experts, and the results of the FGD of the risk assessment (RA) carried out by communities. Community participation in the working group was needed to confirm the experts' work. However, the combination of VCA and RA faced difficulties in addressing different scales and methods. The community relied on a trial-and-error process while the experts and government depended on the risk assessment method.

When asked about the scale of VCA and RA, Pramita Harijadi mentioned factoring the human dimension into the assessment method. She said that the experts seldom think about the human dimension. Moreover, she found in the field that many local people adapt by trial and error:

> It would be more interesting to make a vulnerability assessment if you used one criteria that is seldom used…the human dimension…. If social factors [links to] institution, economy [links to] income…the but human dimension?…. [I]t may be survival capacity and permissiveness… they just accept it in any case….For example, in the flooded area, one family still stays and lives there even though the floor is frequently inundated and his house is half-left…. When we asked the children whether anyone still lived there, they said yes…the parent live there, and they stay temporarily with their friends…. So, the question now, is it an adaptive capacity or a forced situation? Or something else? It is interesting to examine. (Pramita, 26 June 2012)

Unlike the Mercy Corps, which places flooding in the framework of the integration of climate change adaptation (CCA) and disaster risk reduction (DRR), *Action Contre la Faim* (ACF) puts it in the context of disaster management ACF operated a community-based disaster Risk Management (CDRM) program in Jakarta in 2003 and 2004 and from 2007 to 2009. In the latter period, ACF focused on community empowerment and local actors through the strengthening of capacity in the preparedness and response to flood emergencies. ACF also consolidated the integrated local management of floods. According to Rama Furry, a spokesperson for ACF, the people living in North Jakarta's flood-prone areas do not know when flooding might occur. He argued that the Met Office (BMG) has frequently released flood warnings, but the residents of Muara Baru do not have access to them (*Jakarta Post* 2009). Thus, it is important to strengthen the local system.

ACF has worked closely with the *Satuan Perlindungan Masyarakat Penanggulangan Bencana dan Pengungsi* (SATLINMAS PBP), which is the government protection unit for disaster management and refugees. According to the Letter of Decree of Governor of DKI Jakarta Province No. 96 Year 2002, *SATLINMAS PBP* is a joint organization of *kelurahan*, the lowest level of government unit, and a community stakeholder in disaster management. According to Andre Napitupulu, a humanitarian specialist who worked for ACF in Penjaringan, the first program focused on facilitating the development of flood prevention facilities, but in the second phase, they changed the program setting to strengthen the organizational capacity of SATLINMAS PBP. The CDRM program of ACF used participatory rapid assessment (PRA) through community needs identification, community mapping, and calendar mapping in order to identify the kind of capacity building and training activities that were needed by the community, especially vulnerable people.

As an NGO, ACF has intervened in the community organization by establishing functional units. One is *Tagana* (the Indonesian acronym for the youth organization for disaster response). ACF recruited members of this community youth organization; trained them in quarantine and disaster emergence response skills, such as search and rescue; and encouraged them to train their colleagues. This organic unit is expected to be in continuous operation and to preserve the knowledge and skills learned from ACF. However, based on the story of Mr. Napitupulu, the project officer of ACF, I interpreted that the program has not been fully successful because it stopped the implementation without monitoring and evaluating the programs. He was not even sure that *Tagana* was still active. This example indicated that introducing a new program into local institutions requires preparation time and control tools.

In his work on CDRM, Andre was convinced that the community already knew that they lived on the floodplain and a risky riverbank. The residents knew that the flood only lasts for 6 days a year, so they still had almost 360 flood-free days. They also knew that they could expect assistance during the flood. He added that the head of the *kampung* was the liaison between external parties and the community. This person had access to government facilities, launched the strategic planning process, and rallied participation in public meetings or activities. Therefore, the government and NGO recognized the importance of heads of neighborhood associations in policy and program implementation (interview with the author, 06 October 2012).

Another worthwhile micro-adaptation program is the Planning for Integrated Coastal Adaptation Strategies (PICAS) for North Jakarta security, which was organized by the Indonesian Association of Planners (IAP). In 2012, IAP conducted qualitative community-based adaptation planning in Kampung Kamal Muara (KKM) and Kampung Kebon Bawang (KKB). IAP used critical cause analysis to identify the flood problem and conducted participatory planning through a series of focus group discussions in order to identify the community's adaptation plan. According to the project manager of PICAS, Dhani Muttaqin, IAP is concerned with community adaptation because the poor settlers in North Jakarta are the most vulnerable to climate change because of their low incomes and lack of education. Therefore, he argued that the self-empowerment of these people should be through capacity building (interview with the author, 09 September 2012).

Another planner who worked on the PICAS project, Raka Suryandaru, insisted that the impact of climate change would only worsen the social problems of the most vulnerable groups. He thought that community participation could lead to new insights and data to adaptation planning. The PICAS project recognized that the best solutions for disasters related to climate change would come from the communities. However, those community solutions will need to be incorporated into the city development and spatial planning. He argued that risk management performed by the people of *Kampung Kamal Muara* (KKM) and *Kebon Bawang* (KKB) would be sustained only if the government accommodated those activities. Therefore, PICAS provided a rationale for the integration of the activities into formal spatial and development plans (Raka Suryandaru, interview, 20 October 2012).

Based on several community initiatives in Jakarta, the NGOs recognized that the "climate proofing" conducted by experts might not offer convincing data or reliable climate projection. At the micro level, the experiences of people should not be used merely to confirm or validate vulnerability assessments but also to be should be integrated into the analytical process. The NGOs realized that people's experiences in adapting to climate change comprise a key variable at the community level. The ACF's case study found that the adaptation plan could not be institutionalized because the NGO brought new knowledge that required time for the community to internalize, even though it had already adopted participatory planning. According to Sagala and Damayanti (2010), most community-owned initiatives were not communicated through planning processes. In other words, the value of adaptation planning at the community level is yet recognized.

3.4 Different Types of Adaptation Planning

Theoretically, there are three options in adapting to floods: protecting the area, relocating the settlement, and living with the hazard. Based on the planning initiatives and practices in Jakarta, it seems that from the Dutch colonial period onwards, the objective of the government has been to protect the coastal area. The planning has gradually evolved in response to the current situation and future challenges. The government is expected to continue protecting the city from floods by developing flood infrastructure. However, other stakeholders have chosen different objectives while they await the results of the government's adaptation plan.

This research found that the domains of the planning process differ. These various departure points have created a gap that could be overcome by integrating both perspectives into the adaptation planning process. The background of the actors determines the form of the space where the dialogue about planning ideas and concepts takes place. At the micro level, most of the discourses center on living with floods, without considering that the flooding could be disappeared. At the macro level, the discourse is focused on how the flooding could be reduced and how the city could be protected.

The second divergence was found in the adaptation planning process itself. Numerous studies and practices in Jakarta reveal two types of processes. The first type, community-based planning, emphasizes the interaction of experts and community; the second type depends on experts to assess the vulnerability and select adaptation options. The two types have different drivers, actors, goals, data, methods, and knowledge (see Table 3.6). The first type is associated with NGOs, whereas

Table 3.6 Typology of processes of adaptation planning in Jakarta

Elements of planning	Type 1 community-based	Type 2 climate proofing
Drivers/actors	Facilitators of CSO/NGO in CCA, DRR, and/or sustainable development	City planner/professional who has a competence on CCA and/or DRR
Goals	Increasing coping capacity to flood	Reducing vulnerability to flood
Data	Climatic and non-climatic data/information	Climatic and non-climatic data/information, and experts' justification
Methods/approaches	Participatory planning	Rational comprehensive planning
Types	Prioritized actions	Systematic response
Level	Community to subdistrict level	District/city to national level
Source of knowledge	Modules/guidance	Theories/practices

Source: the author

the second type is associated with government institutions. However, in some cases, the donors have supported both.

The micro perspective sees people's experiences as a source of knowledge; therefore, it depends on traditional and tacit knowledge. This perspective is embodied in the realm of the people who experience the phenomenon. In contrast, the macro perspective considers quantitative analysis and expert opinion. Transparent and accountable processes, with certain limitations in assessing flood-related vulnerability, are typical of the macro perspective.

Traditional knowledge is usually derived from the experiences of local people who are vulnerable to the effects of climate change. Such knowledge is a key information source for the planning process. Based on several cases of MCI, ACF, and IAP, the communities in Jakarta have adapted to floods. In Vietnam, the people who plant mangroves to prevent storm surges in the Mekong Delta (UNDP 2008) could provide insights into local adaptation planning practices. The Papuans who use their traditional tenure management system, known as *sasi,* conserve local biodiversity (McLean et al. 2009). Similarly, the people in Toineke Village, who have adaptation strategies that are embedded in traditional practices and local knowledge, support their livelihoods (Hornidge and Scholtes 2012). If all the mechanisms of information sharing among residents' planned adaptation practices are socially constructed, lessons could be learned. Laukkonen et al. (2009) recommended that the development of a methodology and tools to help individual residents and their communities in the planning process is required not only to gain their participation but also to embed their knowledge and actions in adaptation planning.

The use of traditional knowledge in adaptation planning can be successful through building connections between informal communities and formal institu-

3.4 Different Types of Adaptation Planning

tions, as well as through collaborative planning among stakeholders. To build such connections, planning actors should facilitate the transmission of the knowledge of adaptation planning. Transmission is the way in which knowledge alters local behavior (Carmin et al. 2012). External factors, such as professional networks and associations, NGOs, and consultants, often transmit ideas and knowledge to the local actors in cities. These ideas, however, are not easily transferred to the individuals at the community level because technocratic language must be translated into a language that community residents will understand. In Jakarta, linking community-based adaptation with the urban development plan is challenging. The codes and sectoral approaches of governmental bureaucracy need to be synchronous with the adaptation plan formulated by the *kampung* people.

The adaptions of the community and the Jakarta government to floods rely on different pathways and sources. People of the *kampung* still rely on traditional knowledge and actions, which should be integrated into the governmental processes. According to Rabe (2011, 37), "Indonesia needs to harness its long tradition of community participation and self-help to come up with innovative local solutions to integrate poverty reduction with climate change actions." He suggested that engaging the community in an early stage generates relevant information and knowledge. Moreover, the capacity building of the vulnerable group is central in adaptation planning. Without the disclosure of the community's adaptation planning, there will be no connections among planning outputs. Nonetheless, the city government has yet to understand these differences and to integrate the community's planning into a city plan. Therefore, institutionalization is difficult.

Several cases of adaptation planning in Jakarta revealed that the divergent worlds of adaptation planning interfere with institutionalization. The different realms of planning produce inefficient and ineffective adaptations and potentially create new problems. Climate proofing does not solve the problems at the community level, and community adaptation planning is not recognized at the city level. The NGO's involvement in participatory planning does not bring experts and vulnerable groups together. It makes up for the existence of those realms at different levels but not as structurally connected frames. These realms will still produce ineffective adaptation if the local institutions cannot bridge the knowledge gap among the actors. Not only scenario planning based on climate proofing but also the lived experiences of vulnerable groups should be integrated with city development planning. The bridging processes will streamline the institutionalization of adaptation planning, which will eventually lead to strong institutions of adaptation to climate change.

These processes underline the emergence of people-centered planning. This kind of adaptation planning should rely on the local knowledge embedded in each community. The shift in focus in the planning process from climate proofing to locally embedded knowledge of vulnerable people will facilitate the connection between the macro and micro perspectives on flood. Therefore, planning institutions should

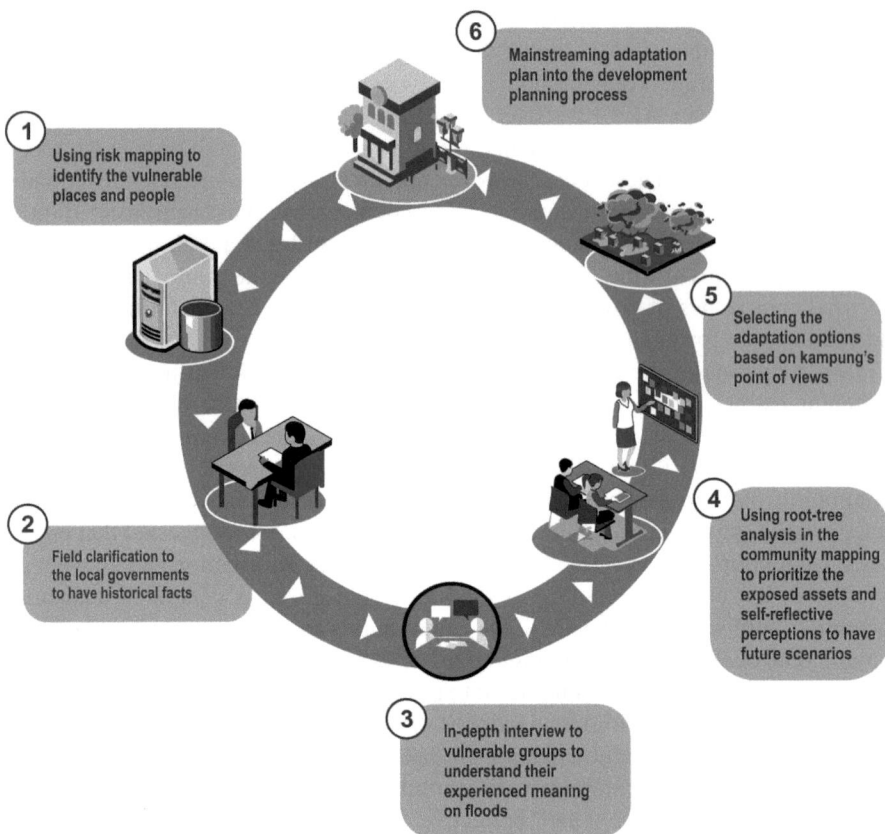

Fig. 3.10 The mainstreaming process of adaptation planning from community to city level (Source: the author 2013, based on IAP (2012))

provide an enabling environment for convergent and loop connections (see Fig. 3.10). Such an environment could be formed through collaboration among stakeholders to ensure that in the future, planning will belong to the people, and it will be implemented by them. It cannot be institutionalized in a single process of adaptation planning that is limited to experts and governments.

Chapter 4
Flood Experiences: "The Vulnerable" and "The Adapter"

Abstract This chapter presents the context where interplay between the floods and the poor takes place. It is an analytical description of the nexus of the urban poor and the floods that are perceived as a major problem in urban development both globally and locally. Limited assets, the lack of access to resources and powers, and the low economic profile make the poor vulnerable. However, many of the poor continue to live in the floodplain. Their perceptions of the stresses and shock caused by flooding could explain their adaptation. Here, I apply *lifeworld* analysis to investigate the construction of the *Kampung Muara Baru* (KMB) people's perceptions of floods. It continues by discussing flood-related vulnerability based on the perception of the KMB people. I then use this construction to discuss the concept of flood-related vulnerability and to identify "the vulnerable" and "the adapters."

Keywords Vulnerability • Kampung • Floods • Stress • Adapter • Vulnerable

This chapter presents the context where interplay between the floods and the poor takes place. It is an analytical description of the nexus of the urban poor and the floods that are perceived as a major problem in urban development both globally and locally. Limited assets, the lack of access to resources and powers, and the low economic profile make the poor vulnerable. However, many of the poor continue to live in the floodplain. Their perceptions of the stresses and shock caused by flooding could explain their adaptation. Here, I apply *lifeworld* analysis to investigate the construction of the *Kampung Muara Baru* (KMB) people's perceptions of floods. It continues by discussing flood-related vulnerability based on the perception of the KMB people. I then use this construction to discuss the concept of flood-related vulnerability and to identify "the vulnerable" and "the adapters."

4.1 Kampung Muara Baru (KMB): The Interplay Between the Floods and the Poor

As mentioned before (Chap. 3), in the North Coastal Jakarta area, one of the most probable districts to experience flood is *Kecamatan* Penjaringan. Bordering the Java Sea in the north, Pluit Lake in the west, Jakarta Bay in the east, and a highway toll road in the south (see Fig. 4.1), *Kecamatan* Penjaringan consists of five *kelurahans*,[1] which are Penjaringan, Pluit, Pejagalan, Kapuk Muara, and Kamal Muara. The total area is about 395.43 ha, and the population is about 306,456, which is located in 65 RWs[2,3] and 800 RTs (BPS Provinsi DKI Jakarta 2013). The development of Jakarta Fishing Port, which started in the 1980s, was the tipping point in the urbanization of *Kecamatan* Penjaringan. Begun in the sixteenth century in the era of the Dutch colonization, *kecamatan* Penjaringan has been affected continuously by new urban infrastructure. The development of water-draining channels, canals, and water reservoirs crisscross the area, such as Cengkareng Drain, Pluit Reservoir, and dykes along the shoreline, has attracted housing properties, such as *Pantai Mutiara* resorts, *Pluit* residences and mall, and apartments. However, there is no low-income housing. Consequently, the urban poor build nonpermanent houses next to the water facilities.

Kelurahan Penjaringan[4] has the largest number of poor households (Firman et al. 2011) and is the most vulnerable to floods (Susandi 2009). Floods have differ-

---

[1] *Kelurahan* is the smallest unit of the Jakarta Province Government based on the Governor Regulation of DKI Jakarta Number 147/2009 regarding the governmental organization of DKI Jakarta Province. *Kelurahan*'s task is to implement the government's duties delegated by the Governor and to coordinate government duties in the area. Its functions include the maintenance and development of public facilities and health facilities in the *kelurahan*, monitor rental housing and green open spaces, and empower RWs (association of RTs) and RTs (neighborhood association).

[2] Rukun Warga (RW) is a neighborhood association that consists of several Rukun Tetanggas (RTs). The RT is the smallest neighborhood unit in the administration settlement in Indonesia. The head of RW or the head of RTs is a documented resident who is selected by residents through an election mechanism every 3 years.

[3] RW administratively represents an "urban locality group" (Dwianto 2003, 4). Although the *kampung* is not considered an official administrative unit, the identification of *kampung* can be approached by the presence of RWs. According to the Governor Regulation of DKI Jakarta Province Number 36/2001, the RT/RW has rights and duties, such as promoting integration between the people and the government, accepting and implementing all of the government's efforts and plans for the development of society, preserving and promoting the Indonesian people's spirit of *musyawarah mufakat* (mutual consultation) and *gotong royong* (mutual assistance), collecting dues, and making full use of any means available for the improvement of the living conditions of the people, etc.

[4] Kelurahan Penjaringan is located below sea level (−1 m) and is crossed by three rivers that flow into the Java Sea: Kali Ciliwung, Kali Angke, and Kali Krukut. The area is about 395 ha and is divided into 17 RWs and 255 RTs. In 2010, the population was about 79,399 with a birth rate of 45 people per month and a migration rate of 27 people per month (BPS Provinsi DKI Jakarta 2012). The coastal condition of Kelurahan Penjaringan causes some RWs, such as RW 01, 02, 03, and 17, to be inundated during rainfall and/or the high-tide season (JICA 2011).

4.1 Kampung Muara Baru (KMB): The Interplay Between the Floods and the Poor

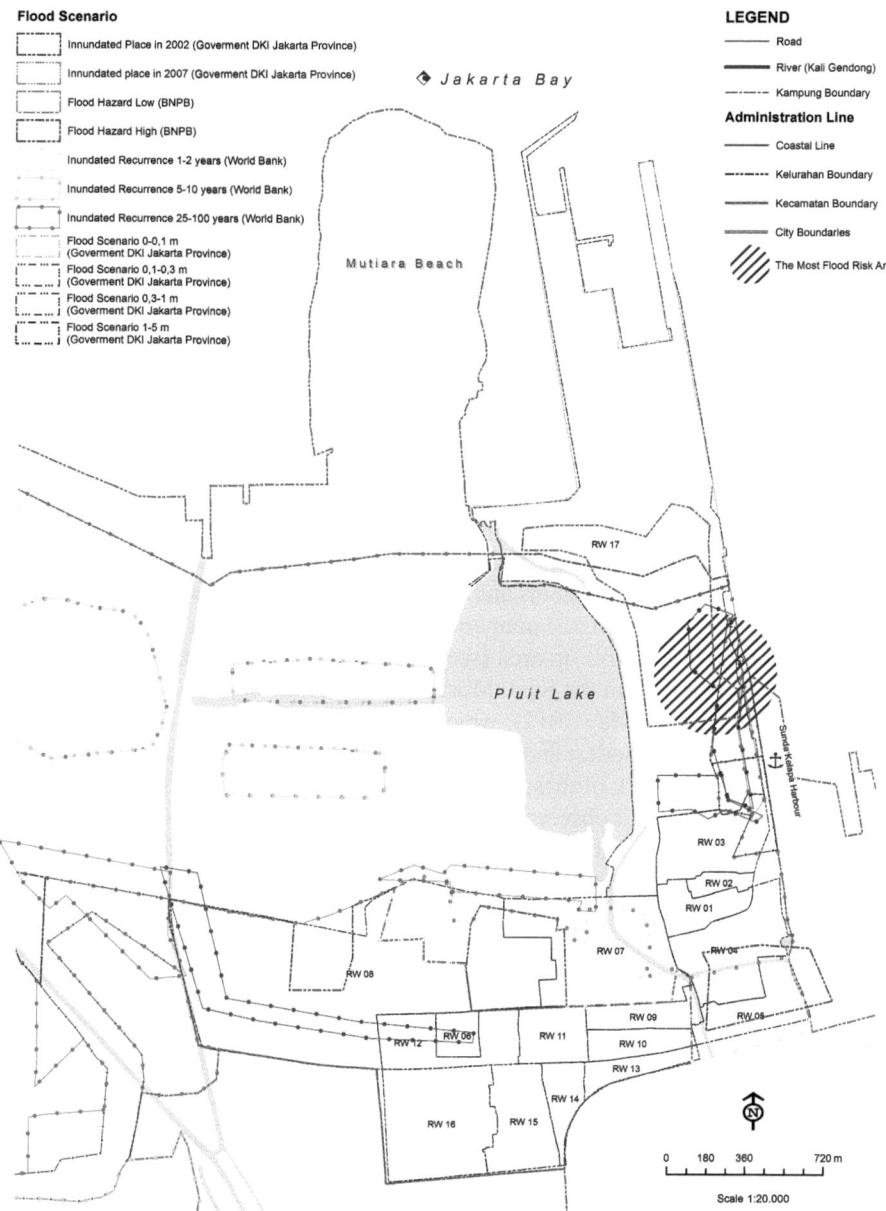

Fig. 4.1 The result of superimposed analysis (Source: the author)

Table 4.1 Number and types of buildings in Penjaringan

No	Types of buildings	Total
1.	Permanent	5781 units
2.	Semipermanent	7569 units
3.	**Slums**	**4574 units**
4.	**Squatters above the river/lake**	**3560 units**
5.	Squatters bellow the toll road	–
6.	Apartments	2 blocks
7.	Flats	8 blocks
8.	Public houses	2 units
9.	Shops/stores	425 blocks
10.	Mall	1 unit
11.	Warehouses	145 units
12.	Industries	266 units

Source: Annual Report of *Kelurahan* Penjaringan, 2011

ent effects on permanent houses and nonpermanent houses. The permanent settlements have good drainage systems and the elevation of their houses higher than that of the local road. In contrast, the nonpermanent houses, such as in the *kampung,* that are predominant in the built-up area (see Table 4.1) lack drainage, and their elevation tends to be lower than the road. Most of these houses are small and attached to each other (Field note, May 2012), which creates the impression of the density of these settlements. The modest and dense construction with poor sanitation and sewage has increased the risk of these buildings and their inhabitants.

According to my field observations, most of the *kampung*s are clustered on riverbanks, along the shoreline, and between industrial blocks. The residents have little or no access to clean water. Because there is no space for wells, and the rivers and lakes are contaminated, water is delivered in hydrant cars. The water company covers only 80% of the *kelurahan* and mainly serves permanent houses and buildings. However, the *kampung* people do have access to education, health, and other social services. In addition, *Kelurahan Penjaringan* has two offices that coordinate disaster management and one station that monitors the floods (see Appendix 6). Hence, the government of DKI Jakarta Province is prepared to manage floods at the *kelurahan* level.

As mentioned in the introduction, most planning practices have been conducted from the *kelurahan* level to the city level, and most studies of adaptation to climate change (CCA) and disaster risk reduction (DRR) have been based on climate impact modeling, social economic vulnerability, and floor risk assessment. Their findings showed that the interplay between floods and the poor occurs at the macro and mezzo levels and increases flood-related vulnerability. However, nothing is known about how the *kampung* people perceive the interplay between floods and poor in their own *kampung*. It is important because the *kampung* and *kampung* people represent the localities in the urbanization of Jakarta. Combining the arguments of three scholars, I define a *kampung* as a "tight agglomeration of continuous and

4.1 Kampung Muara Baru (KMB): The Interplay Between the Floods and the Poor

incrementally developed self-help housing" (Silas 1990, 45), "an indigenous and unplanned settlement" (Silver 2008, 130), and an informal, accretive, and unserviced urban village that houses a majority of the urban population (Garr 1989, 79). Therefore, the interplay between floods and the poor should be examined in relation to the *kampung* of Jakarta.

However, according to the ACF flood risk assessment,[5] which used the five variables of elevation (hazard variable), the seawall and water pump installation (capacity), the distance from the sea or river, and the building material of the houses (susceptibility), RW 01, 03, 04, and 17, possessed high flood risk and endangered the lives of 16,488 inhabitants or 30% of the total population (ACF 2008). *Kampung* people are vulnerable to injury, property, and income loss because they live and work in the low-lying coast and along waterways. The water scarcity caused by saline intrusion is another problem caused by floods. Furthermore, a flood simulation of North Jakarta indicated that the *Kelurahan Penjaringan* is one of the potential areas that will be inundated by 2050 (Hadi 2010). According to the World Bank (2011, 33), "the combination of these hazards and the economic and physical fragility of the poor put them at high risk for loss of property, illness, economic disenfranchisement, social disruption and displacement." Therefore, the *kampungs* in *Kelurahan Penjaringan* are at risk.

I used five available maps to identify the *kampungs* that best showed the interplay between the poor and floods. The first shows the distribution of flooded areas in 2002 and 2007 on the flood map issued by the government of DKI Jakarta Province. The second map showed flood threats and hazards based on historical data of BNPB. The third map showed a potential flood scenario in 2030, which was issued by the government of DKI Jakarta Province. The fourth map was issued by the World Bank Indonesia and showed the recurrent intervals of flood inundation and the probability of floods in Kelurahan Penjaringan. The fifth map showed the distribution of poor households in *Kelurahan Penjaringan*. I used the classification system and data compiled by the government of DKI Jakarta Province. These spatial data were superimposed to identify the *kampungs* (RW) that had been impacted by floods and that housed poor residents. The *kampung* that had the most intersectional value was selected. A detailed explanation of the spatial data processing and analysis is provided in Appendix 7.

Based on the results of the overlay process as described in Appendix 7, RW 17, or *Kampung Muara Baru* (KMB), showed the greatest amount of data that intersected floods and the poor (see Fig. 4.1). The other reason for choosing KMB was also its geographical location on the coastline, which best represented the phenomenon of coastal flooding. The presence of several community-based programs was also considered in selecting KMB as the location for phenomenological research.

---

[5] In 2010, the ACF conducted a study of disaster risk assessment in three *kelurahans*: Penjaringan, Kampung Melayu, and Cipinang Besar Utara. With regard to Penjaringan, it stated that three types of disaster risk pertained to Kelurahan Penjaringan: floods, which potentially impacted 16,488 inhabitants or about 30% of 55,780 total population in 5 RWs from 17 RWs; fire, which potentially impacted 50,722 inhabitants or about 91% of total population in 15 RWs; and dengue, which potentially impacted 33,508 inhabitants of about 60% of total population in 10 RWs (ACF 2010).

Fig. 4.2 Growth of KMB population (Source: Kelurahan Penjaringan (2012))

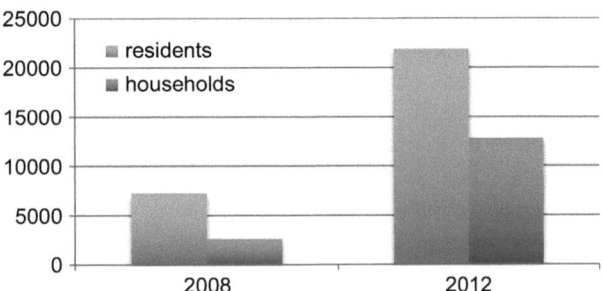

4.2 Factors Affecting the Floods in Kampung Muara Baru

The superimposed analysis showed that floods and the poor intersect in KMB.[6] The profile of KMB will produce a holistic picture of the KMB people who have lived with floods. Their life situation, social dynamics, economic condition, and environmental changes over the last few decades must be understood before their *lifeworld* before exploring their vulnerability, adaptation planning, and adaptation practice. I analyze six factors that must be discussed to understand the KMB people: the characteristics of KMB dwellers, land and building status, basic services, community leadership, and economic condition. The discussion of these circumstances revealed the reasons that KMB is the place where flooding and the poor intersect.

4.2.1 Types of KMB Dwellers

The documented population of KMB is about 21,865 inhabitants or 12,818 households; 65% of the population are adults and 54.5% are males (Kantor Kelurahan Penjaringan 2012). The residential density is 19,552 people per square kilometer, which is distributed among 22 formal and 28 informal RTs. KMB is the densest *kampung* in Kelurahan Penjaringan; the population of other RWs is just under 10,000 inhabitants. In addition, the KMB population has grown significantly from 2008 to 2013 (see Fig. 4.2). The 300% population growth in 5 years, as well as 602 in-migrants and only 90 out-migrants during February 2013, is an evidence of the attractiveness of KMB (Kantor Kelurahan Penjaringan 2013).

---

[6] KMB is administratively registered as a Rukun Warga (RW 17), which is a part of Kelurahan Penjaringan. KMB is located in the north coastal area of Jakarta and is surrounded by the Java Sea on the north side, the Kali Opak River along the Sunda Kelapa harbor on the east side, the toll road Cengkareng-Pluit and Bandengan Utara on the south side, and the Pluit Reservoir and *Jembatan* Street on the west side (see Fig. 5.6). *Muara Baru* means new mouth of river or new delta, which describes its geographical location as a delta region and its built-up environment as a new settlement. The total area of KMB is only about 1.12 km².

4.2 Factors Affecting the Floods in Kampung Muara Baru

Based on the number of poor households and the amount of informal housing in 2011, the government of DKI Jakarta categorized KMB as a high slum[7] and defined it as the largest slum in Kelurahan Penjaringan (BPS Provinsi DKI Jakarta 2012). Slum households comprised about 96.7% of the households in KMB or 2979 households (Sentosa 2010). Mercy Corps Indonesia (2010) categorized RW 17 (KMB), RW 12, and RW 10 as slum areas in Kelurahan Penjaringan. The government data on the number of *Raskin* (*bahasa* acronym for rice for the poor) also showed that there were 3107 *Raskin* recipients in 2012 (secretariat of RW 17, October 2012, quoted by the author). However, the distribution of *Raskin* recipients, as representation of the poor, is interesting because most reside in the RTs that are located on the lakeside where many undocumented residents live (see Fig. 4.3).

In addition to undocumented residents, there are temporary migrants who do casual labor but are registered in the RTs and live in the modest rental housing of KMB, such as RT 7, RT 15, and RT 20. In the house of the head of RT 15, Agus and his five family members rent three very small rooms. Two people rent each room and work in the fisheries, harbor, and warehouses. Therefore, in Agus' house, there are six documented and six undocumented people. All are poor (Field note, August 2012).

Furthermore, the ethnicities (plural community) are a characteristic of the KMB people. Although 97.6% of the population is Muslim, there are also Betawi, Tegal, Bugis, Sunda, Padang, Madura, and Chinese in the KMB area. These people are identified through their livelihoods. Betawi people usually carry fruits and vegetables; Madura people sell water or become junkmen; Tegal, Padang, and Sunda people sell food; and Bugis people work as fishermen. These segmented occupations have helped with the adaptation to floods. In responding to the emergency following the flood of January 2013, Sunda and Padang women served food in the public kitchen (Field note, January 2013).

In terms of livelihood, most KMB people work in the fisheries and factories (see Fig. 4.4). However, this figure includes only documented residents. Most of their jobs are linked to the fishery business. Only a small number of jobs are in other occupations, which indicate that most permanent residents of KMB depend on the flood-sensitive private sector for their income.

The KMB people, whose community comprises formal and informal residents, multiple ethnic groups, and a mix of livelihoods, show us that in order to understand *kampung* people, we need to know the kinds of people who live there, how long they have lived there, and why they live there. Generalizing the urban poor will not help in understanding their adaptation practices. We should not measure the adaptive capacity of the urban poor by simply asking them the same questions. We need to understand their *lifeworld* so that we can assess their capacity.

---

[7] The UN Habitat—United Nations Human Settlements Programme—defines "slum" as an area that combines inadequate access to safe water, inadequate access to sanitation and other infrastructures, poor structural quality of housing, overcrowding, and insecure residential status (UN-Habitat 2003:12).

Fig. 4.3 Distribution of poor households (Source: the author, based on the secretariat of KMB)

4.2 Factors Affecting the Floods in Kampung Muara Baru 73

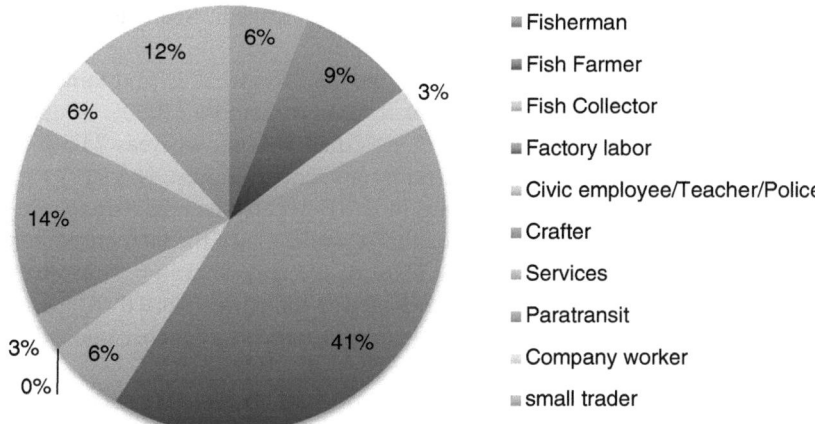

Fig. 4.4 The livelihoods of KMB residents (Source: Secretariat of KMB (RW 17), compiled by the author)

4.2.2 Land and Housing Status

The status of land and housing status influences their adaptation options. Property owners will consider floods differently than renters will. Moreover, floods affect different buildings in different ways. The stronger a building is, the less damage a flood can cause. Therefore, it is important to understand the land and building status of KMB.

Most KMB people do not have land certificates because the land belongs to the state or to private companies. They have occupied the land *en masse* since it was abandoned after the 1997 economic crisis. The national emergency forced the government to let unemployed people rent the "sleeping land" and use it as cultivated land.[8] According to Konedi, the Secretary of RW 17, the land next to the lake, which is occupied by RTs 19, 17, and 16, belongs to BP Pluit, a state-owned company in DKI Jakarta Province, which was ordered to manage the property in the Pluit Lake area. Land next to the sea, such as RTs 21, 15, and 3, belongs to PT Pelindo, a state-owned company that manages labor for the central government. Other lands have

---

[8] According to the regulation of Head of National Land Office Number 02/2003, *tanah garapan* or cultivated land means a piece of land that has already been or has not been attached yet by the certain right which was used or benefited by the others with or without approval by the respected ones with or without a certain length of time. It means the land is not owned, but only managed by individuals and groups through certain occupation rights. If the cultivated land belongs to the state and has been used for decades, as proven by the letter of Lurah (head of subdistricts), the land has utilization rights or even proprietary through the land registration in the city office. However, if it belongs to an individual who already has a title, it cannot be registered. *Tanah garapan* can be originated from the customary land that is not converted to the certain titles of land or from the state land, including the state land that was managed by the state offices and state agencies, or from the wasteland or displaced/dormant land. Despite the land's title, since it was not occupied or used, the local people or migrants occupied those types of land.

been given to the companies for industrial activity (Konedi, Secretary of KMB in discussion with the author, May 2012).

Thousands of households have occupied the land for decades, and turned it into the dense residential area that it is now. According to Sentosa (2010), around 76% of the built area in KMB takes up an area less than 100 m^2, whereas 73.5% of the land ownership is without status and 26.5% are rented houses. The story of land transactions and residential developments in KMB began when the fishing port was developed in the 1980s. Before then the initial dwellers used the land only for small farming activities, especially fishponds. Most of the area is swamp, so it was used only for subsistence farming. When the fisheries industries started to operate in 1990s, the conversion from swampland began, and some dwellers became industrial workers.

As the national economy improved and the job market revived in the early 2000s, the number of workers migrating to KMB increased. They built houses in groups, sometimes telling the landowner but sometimes not. According to "Karyo," a resident of RT 7, he built his house by secretly taking land that was far from the main street (Muara Baru Street) to avoid inspection. He built his shelter gradually with the help of three relatives from his village. A few groups were also building houses at the same time, he said. They initially built only shelters, or *kotak* (box), because of their simplicity and small size (about 3 x 6 m). They used modest materials at first, including cardboard. Then they fixed and upgraded it gradually when they could afford to do so (Field note, June 2012).

The second generation of residents usually bought land or a house by using only a payment receipt. This transaction involves two parties, and it is witnessed by the head of the RT or a neighbor. In some cases, a *girik*[9] letter is appended to the transaction, but in other cases, a letter from the head of RT/RW clarifies that the land has no dispute or other problems and explains the history of the land. For example, Pak Irpan, a resident of RT 15, bought a 3 x 6 meter (m) house from his neighbor to build a *bedeng* (*a* temporary house or shelter). He obtained the notice of transaction, but he did not worry about the land title because according to him, as long as the head of RT knew, it was acceptable. He first modified the house into two rooms and then built the second floor 3 years later. Now he has four rooms that he rents out. Rental houses are common in KMB. Unlike Pak Irpan, Agus, the head of RT 15, added two rooms to rent on the second floor of his house after his father moved back to his village. He rents them out monthly. Both unfurnished rooms are 2 x 2.5 m. According to Agus, in his RT, 12 to 14 houses rent rooms, and there are 4–5 rental houses (Field note, October 2012).

Berner (2000, 7) argued, "transactions in the informal land market are not controlled and registered by the local authorities." The informal submarket has met the high demand for housing in KMB. Therefore, although there is no land title for individuals and building permits for residential uses, the land and housing market is

---

[9] *Girik* is not evidence of land ownership, but of land controlling and tax payment on a parcel of land and building on it (if any). *Girik* can be registered to be a land title with certain conditions. *Girik* can be used for land transactions as long as both parties agree on the condition.

active through the free submarket, which has replaced formal land regulation. However, "the KMB dwellers always pay land and building tax to the government, even though they do not have a certificate" (Interview with Gus Tara, head of KMB, 11 October 2013).

Between 1980 and 2010, the number of neighborhood associations (RT) multiplied. In the 1980s, KMB consisted of only 12 RTs. Ten years later, there were 15 RTs, and 14 years later, there were 22 RTs. Because the number of KMB dwellers has increased quickly, being the head of an RT is in an inconvenient position. Therefore, the RW has had an RT representative (*Perwakilan* RT) since 2000. *Perwakilan* RT is an overpopulated part of an RT (Field note, October 2012). According to Konedi, Secretary of RW 17, *Perwakilan* RT is needed to serve the increasing number of residents in the RTs:

> Perwakilan RT has risen because of the limitation of the head of RT to serve their people. [Can you] imagine, [if no Perwakilan RT], 1 (one) head of RT serves 200–250 household? Meanwhile, a head of RT only has time after office hours; it makes them frustrated. So heads of RT proposed the establishment of a *perwakilan* RT in the RW. (Interview, Konedi, 20 October 2012)

The total number of RTs and Perwakilan RTs is now 48 units. Each is the same size as an RT, that is, twice as large as before. The fast-growing number of *Perwakilan* RTs in KMB has a strong relationship with the informal temporary migrant[10] who are undocumented. According to one participant, "It is a squatter settlement that occupies the land along the lakeside" (Interview with Gus Tara, head of KMB, 11 October 2012). Therefore, Perwakilan RT is not an administrative area that is admitted by the *kelurahan*, but only by the *kampung* or RW.

The extensive increase in housing has occurred informally and tacitly because the residents of KMB cannot afford to build proper housing. They use cheap local materials such as bamboo for the house's pillars, and plastic or inorganic garbage piles up on their reclaimed land. They do not build a disposal site, drinking pipes, or safety electrical ports. They build or extend houses haphazardly. They inform only the head of the RT, and they do not obtain government permits. Although the heads of the RTs manage social affairs in the community, they do not have the educational or technical background to enforce zoning or building codes or to provide advice on housing permits and/or construction.

> I do not have higher education background; I got elected because nobody was nominated in the election. I did not ever and do not want to know about building regulations. The important thing for me is that a new house or repaired house does not disturb neighbors' buildings... (Interview, Gus Tara, 12 August 2012).

---

[10] Informal migrants are illegal residents. The difference between the legal residents of KMB and illegal residents is based on living time and house location. Residents have been in the KMB for more than 10 years. The illegal residents stay more than 5 years and live in the Pluit lakeside (Interview with Gustara, the head of KMB, 12 October 2013).

This situation has resulted in irregular housing and settlement development as well as questionable building safety because it is based on the owner's knowledge. The city government's acceptance of the expansion of KMB housing has been replaced by traditional practices in KMB. Because there are no regulations about the minimum number of tenants in a house or flat, the sanitation standards, density, and other matters concerning housing transactions are managed through oral agreements between neighbors, which are mediated by head of the RT.

Figure 4.5 shows the irregular pattern of KMB housing along the lakeside, coastal area, and roadside. Some are attached to warehouses, but they do not complement each other, and they contrast the industrial and warehouse area. However, in the neighborhood, the houses follow a ribbon pattern along the road or lake. This building pattern places the residents at risk of both flood and fire. Furthermore, the dense settlement cannot be safely evacuated in the case of an emergency.

There are no written rules regarding the ideal population or type of housing in each RT or *Perwakilan* RT. However, the roles of head RTs or *Perwakilan* RT are significant in the management of the population and the housing because visitors or migrants who want to stay more than 1 day or move to KMB must report to them. Therefore, they should be able to manage the increasing number of households. The lack of capacity of RT heads and the limited intervention from the government has plunged the residential development of KMB into chaos. Driven by the Government Regulation Number 36/1998 regarding dormant land control and utilization, people were allowed to occupy the abandoned land. This regulation has weak criteria for what qualifies as dormant land, which has created difficulties in its implementation (Parlindungan 2003; Supriyanto 2010).

Furthermore, no government agencies control housing development. According to "Anto" who is an officer in Kelurahan Penjaringan, the *kelurahan* knows that most of the buildings in KMB have no formal titles or land permits but does not want to prohibit poor people from living there because it is close to where they work. He added that the *kampung* is supposed to have its own mechanism, which the government respects. It is difficult for the *kelurahan* office to intervene in the housing and settlement in the *kampungs*, including KMB.

The ignorance of the head of the RW and the heads of the RTs about the technical aspects of spatial density and housing safety should be blamed for this situation. Both the head of RW and the heads of RTs focus on the administrative requirements of the housing development proposals. Based on my observations, the absence of building permits, the unknown quota for the numbers of houses, the lack of standards for safety and healthy housing, and the ease of building a house are responsible for the severe effects of flood damage.

4.2 Factors Affecting the Floods in Kampung Muara Baru

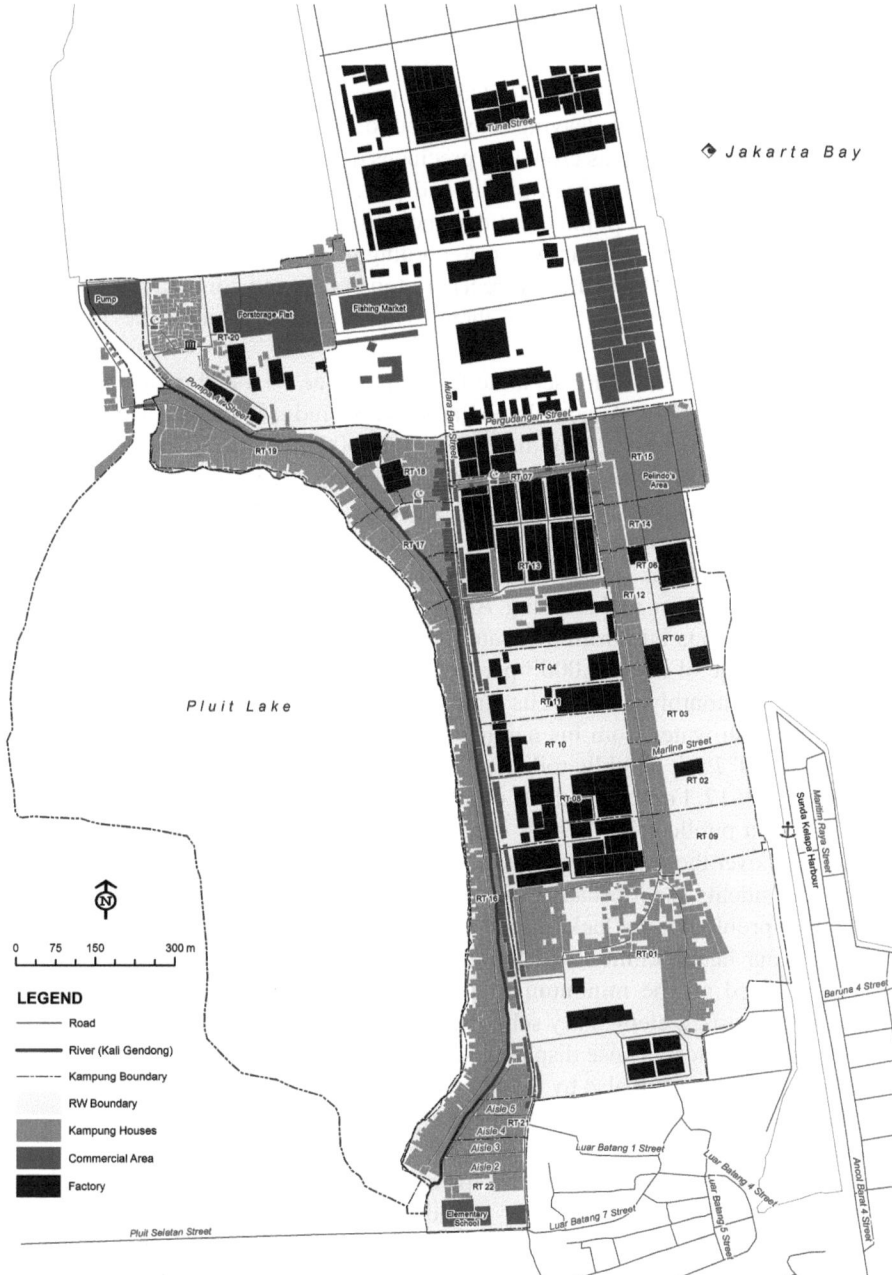

Fig. 4.5 Map of the KMB settlement (Source: the author, based on fieldwork (2012))

4.2.3 Basic Infrastructure

The residential density indicates limited and unstandardized basic infrastructure. Roads are only 1.2 m wide, leaving only 15 cm of drainage that is 30 cm from the front doors of the house. This crowding precludes quick evacuation, endangers children, and blurs the line between public and private (Field note, July 2012). The heads of RT 7 and RT 15 pointed out that the most serious infrastructure problems are drainage, sewage, and clean water. According to them, the open, narrow drains along the main road (Muara Baru Street) are too small, and commercial buildings have already been built on top of them. This situation inhibits the runoff of floodwater.

The drainage system is not separate from the sewer, so a bad odor is produced when the drains are full. Accumulated garbage also reduces the drainage capacity because there is limited space for disposal. For 22,000 inhabitants, there is only one disposal site at RT 21, which is too far for most tenants to reach. Filthy, black water clogs the drains in which children play or scavenge for plastic.

The other problems are clean water and sanitation. There is limited room for wells because seawater infiltrates the groundwater, so the residents have to buy clean water for cooking and drinking. According to Agus, the head of RT 15, almost 80% of the residents buy from the water refill depots. He estimated that it cost his family an average of IDR 23,000[11] per week for his family of six or around 12.5% of his irregular monthly income. His family usually relied on rainwater from a water tank or brackish water from his neighbor's wells for bathing and washing. Many residents of RT 7 use a public toilet for bathing and washing, which is available in each RT in RW 17. For the residents of RT 19, 17, and 16, which are at the lakeside, there is also a public toilet, which is called a *helikopter*: It consists of a roofless wooden box over the river; a hole inside is used as the toilet; the water is taken from the lake. Residents use it because it is a simple and free facility.

Another problem is the lack of social facilities. There are only nine schools, nine mosques, four health clinics, and four doctors' offices for the 22,000 inhabitants. However, based on the minimum standard of basic services in Indonesia, there should be at least 22 elementary schools, 48 mosques/praying places, and 22 community health centers. These disparities are caused not only by the limited amount of land and buildings but also by the limited financial capacity and the culture of the dwellers. When residents get sick, they buy traditional herbal products (*jamu* in *bahasa*) or go to traditional masseuse. For serious illnesses, they go to mini shops or drugstores to buy medicine, follow the advice of family members or neighbors, or rely on what they learn from TV or radio advertisements. If there is an emergency, they will see a doctor. The village culture still influences their preferences.

The lack of open or public spaces is another problem. With more than 15,000 inhabitants per square kilometer, the availability of green spaces and public spaces is mandatory to improve the quality of living and avoid social conflict. There is only

---

[11] IDR 23,000 = 1.5 euro (1 euro = IDR 15,000).

one football field in RT 1 and four vacant lands in RT 3, 11, 20, and 21, all of which belong to private companies or the government. Therefore, the settlement of KMB is vulnerable to fires and floods. The absence of open spaces not only diminishes sociological and environmental quality but also the safety of residents.

The infrastructure of KMB exposes it residents to the hazards of floods. For instance, there is no way for floodwater to be absorbed by the land or to flow into the retention pond. In addition, the infrastructure capacity is not sufficient. Finally, there is no water infrastructure connection between KMB and Kelurahan Penjaringan except access to the social facilities.

4.2.4 Formal and Informal Economies

The growth of the informal settlement in KMB was led by the availability of employment in the fisheries harbor (PPI Samudera Jakarta),[12] factories, and warehouses (see Fig. 5.9). The economy of KMB is affected by the presence of 31 big industries, 245 warehouses, 22 fisheries, 225 small industries, 895 stores, 27 automotive dealers, 285 offices, 12 banks, and 5 gas stations in Kelurahan Penjaringan. These formal economic activities have created a job market that attracts citizens of not only Jakarta but also other parts of Indonesia. Hence, KMB has become a residential target.

Based on a statistical report by the KMB secretariat, adults comprise 66 percent of the total population of KMB. According to Gus Tara, the head of KMB, the livelihood of KMB is derived mainly from informal economic activities, such as traditional market and street vendors, harbor and warehouse workers, and construction workers. According to him, "there are few fishermen nowadays" (Interview, 20 July 2011), which indicates that the economic driver in KMB is in the informal sector. Some experiences of the KMB people are included in Fig. 4.6.

The nexus of the formal economy and the informal economy constitutes the economic structure of KMB and Kelurahan Penjaringan. The formal sector includes the fisheries harbor, trading, industry, and warehouses. The informal sector includes unregulated labor-intensive activities, self-employed entrepreneurs, casual work, small businesses, unregistered activities, and illegal activities such as smuggling. Both sectors play a significant role in building the adaptation pathway. The dependencies between these two sectors also affect the sustainability of the livelihoods of the KMB people.

For example, the fisheries that reclaimed the sea around RT 20 have attracted people from RT 3. According to Hidayat, a fisherman from Makassar, many tenants in RT 3 and its surroundings have moved to RT 20. His family moved to RT 20 when he was a teenager. He described helping his father put their equipment in a *gerobak*

---

[12] It was known as Nizam Zahman Harbor, located in the north of Kampung Muara Baru. It was built in 1984 and then changed to PPS (*bahasa* acronym for the main fishery harbor in Jakarta). It has become one of the five main fishery harbors in Indonesia.

> Amin, 27 years old, a security officer at the industry company, has lived in KMB for eight years. He confessed that he came to KMB without any preparation. He just followed an invitation from his friend, Husni in Pemalang (East Java) who had already worked and lived in KMB. Husni, told him that there was several job vacations in a wood factory, PT Interwood. Then, Amin came along with two friends with the same educational status with him, graduates of junior-high school. He applied for security at first, but was not accepted. But then he was offered to be an office boy, which he took for one and half years. He admitted that he has changed his job at least five times. He just started his work as security since two years ago. He said, "it was fortunate that I live in KMB, there were many jobs here." He also said that there were at least dozens of people in his village who did the same thing as him, going to KMB to find jobs. (Interview, 9 May 2012)
>
> Abdullah told another story. He is a rounded fruit seller who came from the Cirebon five years ago. He was only 19 years old, did not graduate from the senior high school, and came directly to Jakarta because a classmate in junior high told him about job vacancy in the fisheries harbor. But the job only lasted for two months because he did not have strength to do it. He looked for office jobs, but never succeeded. During that period, he worked on daily basis for several warehouses and then followed a fruit seller who coincidentally came from the same city as him. Less than one year later, he sold fruit by himself. His market area is around the KMB daily market. According to him, earning money in KMB is not difficult even though there are few jobs. (Interview, 15 May 2012)

Fig. 4.6 The story of the informal sector

(a cart) and pulling it to the coast. He said that not only his own family but also several of their neighbors moved. Therefore, the nexus of the formal and informal economies is very important in the economic status of KMB.

The livelihoods of the KMB residents, which depend on the formal and informal sectors, contribute to the flood risk. If business ceases during the floods, the residents potentially will have reduced income, which is already tenuous. The informal sectors also depend on the ability of the KMB people to consume goods. Thus, it is important to understand the economic condition of KMB before exploring the *lifeworld* of its people.

4.3 Increasing Flood Risks and Growing Population

KMB is an informal, poor settlement that is characterized by frequent floods because its location is below sea level and it is surrounded by water (the sea, a river, and a lake). The rising sea levels have been predicted to be 0.35 cm per year. However, the elevation of North Jakarta is about 3 m below the level of the sea's surface (Brinkman 2012). Hence, the frequency of flooding will increase. A big flood occurs almost every year, and the level of inundation reaches from 50 to 200 cm, lasting for months in some places (Sentosa 2010). According to the Kelurahan Penjaringan office and

4.3 Increasing Flood Risks and Growing Population

the secretariat of RW 17, KMB is the most vulnerable to tidal floods, and overflows from Pluit Lake occur (Fitrinitia 2011). In January 2013, as many as 8 pumps and 12 portable pumps could not stem the floodwater (Belarminus 2013). Consequently, floods are part of the everyday life in KMB.

Furthermore, KMB is flooded more often than Jakarta is. KMB residents know that when Jakarta is flooded, KMB also will flood. However, when KMB is flooded, Jakarta might or might not be (Field note, December 2012, quoted by the author). Floods often coincide with the full moon. They do not require rainfall, but are caused by rising sea tides in the spring, which are influenced by the moon and the solar gravity of the Earth. The spring tide is called a *rob*. It has probably influenced the perceptions of KMB residents that they live in a flood-prone area.

The settlement history of KMB reveals the role of urban migrants in converting the swamp area into a settlement area, followed by harbor reclamation and industrial development. Nowadays, KMB is one of cheapest destinations for urban migrants who want to work in the capital city of Jakarta. The shifting the physical landscape of the Penjaringan region from "small-quiet fishponds" to "the busiest fisheries industries" has transformed the KMB area from "nothing" to the "densest settlement."

The recent status of KMB has increased its sensitivity to floods as well as the risk of floods. Several factors limit the adaptation of KMB residents to the floods. These include the following: the undocumented residents of KMB have never been included in estimates of flood management, the uncontrolled highly dense housing reduces the area of floodwater and thus reduces the retention capacity of this area, the garbage-clogged drains exacerbate the inundation caused by heavy rains, the water gate and water pump cannot prevent inundation or even overflows from the Ciliwung River, and the informal characteristic of their livelihood plays a daily role. These factors should be considered in understanding the KMB people.

However, the annual floods do not deter people from living in KMB. Since the early 1990s, the number of neighborhood associations and the incidence of floods have both increased (see Fig. 4.7). The population of KMB has also increased, which indicates that floods do not deter settlement.

The interplay between *kampung* and floods in KMB shows that flooding does not constrain or even reduce the spread of houses and settlements for the poor. Instead, it has changed inundations to floods. If no one lived in KMB and if dense settlement did not encroach upon the river, lake, and green open spaces, the floods would be less destructive. The growing settlement of KMB probably increased the flood risk. Therefore, the ability of the 30-year-old KMB settlement to survive and grow should be considered in defining the *lifeworld* of the KMB people.

The status of KMB has demonstrated that one phenomenon must be thoroughly understood. As argued by Hornidge and Antweiler (2012), the localities that are affected by floods occur in different forms and on different scales. Floods have certainly changed the behavior of the KMB people in adapting to floods, but they have neither persuaded them to move out nor discouraged newcomers from moving into KMB. They continue to build their houses and expand the settlements. Despite these disadvantages, the KMB dwellers perceive their world as a place that still

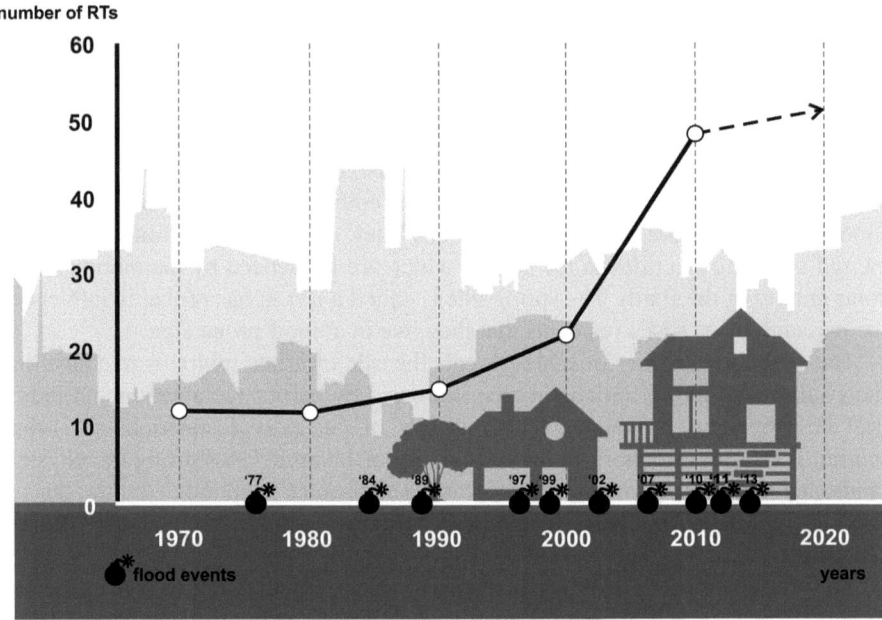

Fig. 4.7 The interplay between poor settlements and floods (Source: the author)

gives them an opportunity to live. Over generations, they have shaped their world with the mutual understanding of individuals and neighborhoods. They have built houses, provided infrastructure, and carried out their daily activities. Therefore, it is important to understand the construction of the world of KMB.

4.4 The *Lifeworld* of the People of Kampung Muara Baru

The perceptions of the KMB people, who live in the flood zone, are interesting because their living place is risky for residential use and economic activities. It seems irrational to live in a flood zone. The unpredictability of floods should be a deterrent to living in KMB. However, the dense settlement by the Pluit Reservoir and along the coastal line is proof that the KMB people are conscious of the potential risks. In other words, they acknowledged the flood risk, but it does not bother them. Therefore, they live in a highly exposed area (floodplain) and a sensitive area (high density) because it is close to their employment. Although they have alternatives in neighboring *kampung*s or inexpensive rental flats provided by the government in the Penjaringan area, they prefer to live in KMB. The longstanding residents are emotionally engaged with their family history, which has accumulated over decades, and newcomers perceive KMB as a strategic place to live.

4.4 The *Lifeworld* of the People of Kampung Muara Baru 83

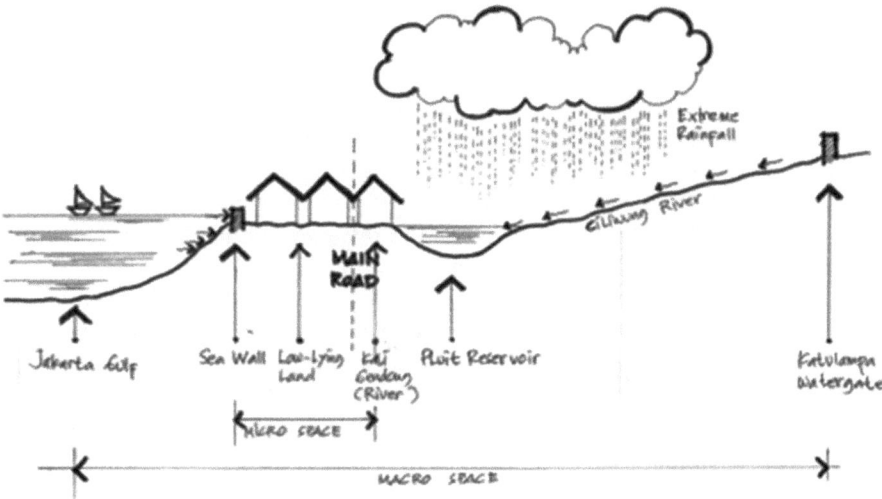

Fig. 4.8 Spatial boundaries of the Kampung Muara Baru people (Source: the author, based on interviews and group discussions in RTs 20, 15, and 7)

In the snowball survey that I conducted in three neighborhood units, RT 7, RT 15, and RT 20, several keywords often emerged when I asked about floods. The interviewees claimed that the flood boundaries were shaped by five spatial elements: *Katulampa Bogor* (Watergate), *Kali Gendong* (river), *Waduk Pluit* (reservoir), *tanggul* (seawall), and *laut* (Jakarta gulf). Katulampa Bogor is located upstream of the Ciliwung watershed that flows into the Pluit Reservoir and the Jakarta Gulf through the Ciliwung River. Kali Gendong is a branch of the Ciliwung River; the sea dyke is a barrier built along the coastal area. The interviewees partially blamed these five components for the floods and emphasized the following:

> Our place is lower than the sea surfaces and it is the end of the water journey from Katulampa Bogor... If the height of the water is more than 180 meter [at Katulampa], we are prepared [for flood]. [The intensity of flood] would be worse if rain comes [whilst] kali [Kali Gendong] and waduk [Pluit reservoir] could not serve. (Konedi, interview with the author, 25 May 2012)

The links among these components are similar to their conception of the watershed; they acknowledged that KMB is the end portion of the water flow from higher land. The water flows through the Ciliwung River and its branches, including Kali Gendong, and enters the Pluit Reservoir before it reaches the sea (Fig. 4.8).

They also linked the components in order to differentiate the types of flooding:

> [There are four types of floods:] *Banjir kiriman* (transferred flood)... from Katulampa through Ciliwung River overloaded in [*waduk*] pluit; *Banjir Rob* (tidal flood) happens when the level of the sea tide is over the sea wall; *Banjir biasa* (rainy flood) caused by heavy rains; and *banjir bandang* (big flood) is combination of all." (Gustara, interview with the author, 11 January 2013)

He emphasized that the ineffectiveness of the Pluit Reservoir and the seawall caused the big floods in KMB. He added that the houses over the river and beside the reservoir also contributed to the floods.

In their *lifeworld*, the boundaries of the flood are framed by the water and the settlement location, which is on the edge of the coastal land. According to Gustara, head of KMB, *Muara Baru* is a new estuary that is positioned below the sea level (Gustara, interview with the author, 18 May 2012). Consequently, the elevation of many buildings has been increased by landfill or by reclaiming land from the sea: "…We also used landfill because our land is slowly sinking without awareness… [It] has occurred for many years" (Kadir, interview, 23 June 2012). In general, "The height of the land fill is based on the level of the highest floods in the past [years] or the elevation level of the main road… but it also depends on a person's budget" (Dulkadi, interview with the author, 4 July 2012).

Agus, the head of RT 15, emphasized the importance of house elevation. He explained that his neighborhood was safer than other neighborhoods around the Pluit Reservoir (i.e., RTs 19, 20, and 16) because it was at a higher elevation. He designated the main road as the marker between the lower and higher land because the left side of the road was around Kali Gendong, so it was originally a riverside land. It was different from his side, where there are many industries. The land was reclaimed during the extensive industrial development in 1990s, so it is higher than the river (*Kali* Gendong). Nonetheless, his place is below sea level and is threatened by tidal floods. The construction of a seawall in 2008 significantly decreased these tidal floods. There was only a small inundation from the permeation of seawater or water overflowing from clogged drains affected by heavy rainfall.

The interviewees initially claimed that the temporal boundaries of the flood are determined by seasonal factors. They agreed that flooding has and would happen at the beginning of the rainy season in *Oktober* (October) and would continue until *Februari* (February). The suffixes "*-ber*" and "*-ri*" are the traditional symbols for the intensive precipitation that is strongly associated with floods. Both words have been transmitted from generation to generation. Since the Dutch colonial period, they have perceived floods as following an annual cycle and a 5-year cycle in the rainy season. The interviewees also thought that the floods followed a 5-year cycle based on the floods at the end of 2002 and 2007 and the beginning of 2013. This reaffirmed the general assumption that there is a major flood every 5 years.

The development of flood infrastructure, from Dutch canals to the present flood infrastructure, which is discussed in Chap. 4, also structures temporal arrangements for flooding. One interviewee expressed the following:

> Before the sea dyke was built, every full moon, we felt unsafe with the high tide, but now we are not worried anymore…[the flood] very rarely visits us again… if it does, it's just a small and short inundation, not a flood…. (Irpan, interview with the author, 27 June 2012)

They sensed that the infrastructure had minimized the intensity and frequency of floods, but they were not certain that the reclamation program would do the same. They also took into account the activity of land subsidence in shaping the duration of the flooding. The presence of flood infrastructure, which has reduced the number

4.4 The *Lifeworld* of the People of Kampung Muara Baru 85

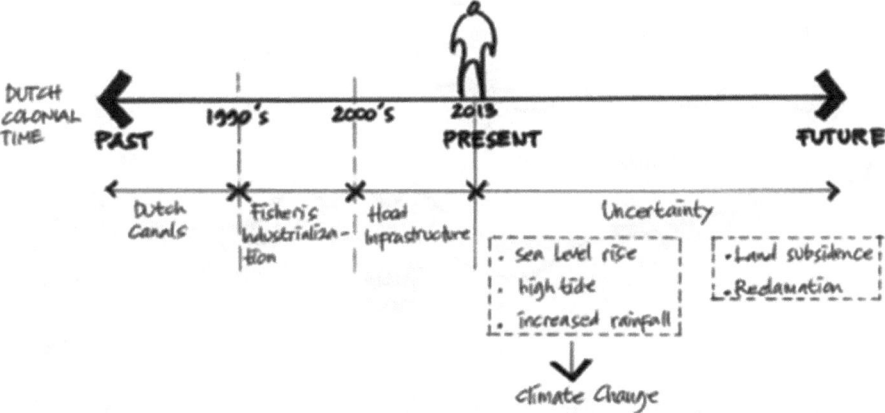

Fig. 4.9 Temporal arrangements of floods in Kampung Muara Baru (Source: the author, based on interviews and group discussions in RTs 20, 15, and 7)

Fig. 4.10 Flooding of KMB on 17 January 2013 (Source: photo taken by author)

of repetitions and the duration of tidal floods, has framed the temporal world of the KMB people (see Fig. 4.9).

Based on their experiences, KMB residents could predict floods. They cited land subsidence, natural change, land reclamation, and flood infrastructure development as factors that could affect the timing of floods. They did not intend to forecast that there would be no floods because they realized that their situation depended on the floodplain and on climate change. However, the flood in January 2013 (see Fig. 4.10) raised questions about the reliability of the infrastructure to handle future flood:

Fig. 4.11 Fragile sea dyke in KMB. (**a**) Eroded. (**b**) Perforated (Source: photo collection of RW 17 (2010))

> We never expected the flood on 17 [January 2013] would be huge and would last for five days. It really shocked us. The water came from the Pluit Lake, which overflowed. It slowly came at night…several groups from RT 20 came to our place and the fisheries harbor. We just stayed for three days without going out from our place… [I]t's like a flashback from the 2007 flood. (Agus, interview with the author, 4 July 2012)

They did not know what a flood would occur despite help from the government. They compared the last three major floods:

> Nowadays [flooding] is uncertain. In 2002, they came in February; in 2007, they came in November; and in 2013, they came in January. A long time ago, [we knew floods would come] every December, but now, [the time is] unclear; [it] could be after or before that… We don't have any clues at this moment; [instead] the Meteorology office (BMKG) could not predict the exact day of this year. (Khairuddin, interview with the author, 5 July 2012)

KMB people realize that the flood infrastructure cannot protect them, especially from high tides in the monsoon season. According to Konedi, Secretary of RW 17, the sea dyke that had been built in 2007 was too low and too thin. He explained that several people in RT 20 had told him that during the tidal flood, the sea surface had reached the top of the sea dyke, and he had found cracks in it (Fig. 4.11) (Konedi, interview with the author, 14 June 2012). In addition to the quality of the sea dyke and the extreme sea tide, the collision of the fishing boats also weakened the sea dyke, especially near RT 20. Hence, the fragility of the sea dyke has increased the KMB people's sense of uncertainty.

The residents also know that nature is continuously changing. The content analysis of the interview data showed that they described natural change in terms of the changing time of rainfall, bigger tidal waves, and the rising sea surface: "When I was child…we saw many flying white ants dying on our floor as an indicator that the rainy season come, but now it's very rare… So, it's difficult to predict the rainy season now" (Eva, interview with the author, 7 October 2012). "Seawater is rising…it appears that our mooring pole is sinking more and more" (Ujang, interview with the author, 16 October 2012). "A long time ago, the sea level is as the same as us.... Nowadays, it seems higher...we are likely underneath [sea surface], especially when it's a full moon" (Mulyono, interview with the author, 6 November, 2012). These factors are linked to climate change, but my informants preferred to use the idea of natural change.

KMB people share their experiences and conceptions in many informal meetings, such as the get-togethers of housewives (*arisan*), *maghrib* discussions among the men, and conversations in youth groups. Housewives have a key role in generating knowledge about floods because they are "the house managers" when floods occur. They know what do to and how to ask neighbors for assistance. The *maghrib* discussion takes place each evening around 6:00 p.m. The men gather before and sometimes after the Muslim prayer time (*maghrib*). Young people gather in the street, and they usually talk while having a light snack that is sold by street vendors, such as *bakso*, *gorengan*, and *buah*. The social relationships formed by these three habits have helped shape the popular perceptions of floods.

Fadilah, who moved to KMB in 2009, is a 32-year-old housewife in RT 15. She told me that she had never lived in a flood-prone area. In her hometown, Rangkas Bitung, 2 h from Jakarta, her *kampung* is located inland. She explained that she has a good relationship with her neighbors and they enjoy visiting each other. She first heard about floods from her neighbor, *Bu Ijah*, who had lived there for more than 12 years. *Bu Ijah* often talked about the floods of 2007 and 2002, including their origins, effects, and the reactions of Bu Ijah and her neighborhood association. Fadilah also heard stories about the floods from other neighbors in *arisan* (social gathering) and *pengajian* (Muslim prayer group) programs. In January 2013 when the big floods occurred, she was frightened because they were worse than she had expected, but she remembered how she was supposed to adapt (Field note, June 2012; Fadilah, interview with the author, 20 January 2013).

The stories about floods in KMB, which Fadilah had never experienced, helped her to tolerate the impact of the flood and to accept the potential losses or damages. Since then, she has believed that the flood would never disappear. Fadilah's reality has been constructed through her existence in KMB and her interactions with neighbors since she started to live there. It is a reality of everyday life in KMB, and she and other KMB residents regularly sense and experience floods. She derived the meaning of flood through her social encounter with Bu Ijah in her *lifeworld*. The face-to-face situations shape ideas about the floods that that new tenants, such as Fadilah, will experience in KMB. These situations encourage reification of what they have shared with each other. *Bu Ijah*'s stories about flood experiences, which were articulated by Fadilah, gave her a benchmark for the meaning of floods.

Furthermore, elderly residents, such as Konedi, Khairuddin, and Agus, who spent their childhoods in KMB, perceived the meaning of the flood as handed down by their predecessors.

They all mentioned the roles of their parents and *sesepuh* (*kampung* elders) in recounting the flood history before they experienced it themselves. Therefore, in addition to their own experiences, they also reified what their parents had told them, and they shared these stories with their children and others in their neighborhoods.

> *Mas* Anas, 23 years old, a warehouse worker, has resided in KMB for six years. In the first year, he stayed in a friend's room. Now he rents a room alone in Agus' house, 500 meters from his office. He has been in more than ten floods, but for him the worst one was the 2007 flood. Having survived the January flood, which forced him to stay on the roof for two days, he did not seem depressed in expressing his experiences: "*Alhamdulillah*, the water was finally dismissed. I thought it would be a week.... It [January flood] was beyond belief.... But, it's returning to normal now...we can work as usual" (Anas, interview with the author, 6 October 2012; 19 January 2013).
>
> *Ibu* Rohima, 48 years, is a housewife with three children and has lived in KMB (RT 07) for 18 years. She moved to KMB because her husband could work in the fisheries harbor and she could run a small canteen (*warung nasi*), which has given them enough income. She is aware of the living conditions in KMB: "There are bad and good things about living in KMB...[bad] floods often hit in the rainy season and fire hazards in dry season....The good ones are [KMB] is close to everywhere and it is easy to get money [work].... Many people come [stay] to KMB over and over even though it's often flooded.... That's why we also do not have any interest to move..." (Rohima, interview with the author, 15 November 2012).

Fig. 4.12 Intentional decisions to live in KMB

They were sometimes surprised that children kept playing during floods. These routine interactions also influenced the meaning that they assigned to flooding. This finding supports Schütz's argument that, in addition to social encounters, the lived experiences of predecessors and successors play significant roles in shaping the structure of the *lifeworld* (Schütz and Luckmann 1974; Berger and Luckmann 1967).

The *lifeworld* of the KMB people indicates that they accept floods as part of life. They knew that the place would continue to be frequently flooded, but they stayed because they could earn a living there. Moreover, their income outweighed the risks. Therefore, they lived there and accepted the flood risk (see Fig. 4.12). The changing spatial and temporal boundaries of floods and the social encounters with others who have had the same experiences of flooding over time have led them to perceive floods as a reality of their everyday lives.

I now discuss what floods mean to the residents of KMB. Many international and government agencies know that in the coast of North Jakarta, especially in Penjaringan where KMB is located, floods are attributable to hydrometeorological factors as well as the effects of climate change, such as heavy rains, rising sea levels, and high tides (BRKP 2009; World Bank 2011; JICA 2011; BNPB 2012). These agencies explained that floods strike KMB because of the failure of the hydrometeorological system in Jakarta's watershed area. Floods are exacerbated by the uncontrolled urbanization following the intensive use of land water and groundwater. The different methods and scales of data have resulted in different explanations for flooding and the locations of flooded areas.

However, the interview data showed that in KMB, floods were perceived differently (see Table 4.2). Most of their terminology implied a lack of concern about the floods and the attitude that they were not out of the ordinary. The phrase *sudah dari sononya* (is given) implies that the flood is a natural phenomenon. In this context, the words *gak apa-apa* (no problem) and *kita sudah biasa* (we are used to it) expressed surrender to the floods, which are out of their control. A flood is a natural phenomenon that occurs because of the Almighty God's power.

4.4 The *Lifeworld* of the People of Kampung Muara Baru

Table 4.2 Perceptions of the floods

The kampung language	Translation	Interviewee
Kita *mah* sudah biasa	We get used to…	Aris, 30 years old, RT 20
Biasa [s]aja sih…udah gak heran	It's a usual (event)… it's not surprising anymore	Kadir, 50 years old, RT 7
Kalau dulu iya, sekarang sudah gak [a]pa-[a]pa	(It was a problem in) the past, now it's all right	Wahyudin, 42 years old, RT 16
Kalau [banjir] datang, ya.. kita gotong lagi ke atas… tapi sekarang jarang banjir *gede*…	When (flood) comes, we bring (our things) up again… but the big flood seldom occurs recently	Ibu Maryam, 52 years old, RT 20
Kalau orang baru mungkin kaget ya…[tapi] kalau kita biasa [s]aja	If new people, yes maybe they have shock…[but] we are just fine	Abdullah Rustang, 40 years old, RT 18
Kejadian banjir memang sudah dari *sononya*	Flooding event is given event…	Ghufron, 25 years old, RT 17
Bukan muara baru, kalau gak banjir	(It's) not muara baru, if it does not have flood	Komarudin, 55 years old, RT 15

Source: the author, based on fieldwork (2012)

After living through several floods, they have accepted that they transcend their operational zone. The KMB people have produced a shared meaning of the flood as a natural event. This belief is influenced by their religious and cultural values. The religious meaning of the flood, which is based on Islam, has influenced the perceptions of the KMB people. Muslim residents have a strong institution at either the neighborhood level or the *kampung* level, which facilitates the transmission of the religious meaning of disasters to the adults in the mosque, women's prayer groups, and children through education in formal and informal Islamic schools. In Islam, Muslims believe that the floods are caused by human disobedience to God (Makmur 2014, translated by the author) and "In the earth, the flood can be found as a sign from God for people to maintain the natural environment as stated in QS Adz-Dzariyat [51]: 20" (Zein 2013, translated by the author). Because people have mistreated themselves (e.g., by building houses in the upstream area and throwing garbage into the rivers and drainage systems), the environment cannot protect them from the floodwaters. The religious reason for the flooding event has relieved the KMB people from their sins by undergoing the floods.

Javanese culture also has shaped the meaning of floods. Most Javanese are street vendors and fishermen. They come from provinces in East and Central Java. According to their calendar, of the 12 months in a year, it rains for three: *kanem* (the sixth), *kapitu* (the seventh), and *kawolu* (the eighth) (see Fig. 4.13). The rainiest period, *kapitu*, is from 22 December to 3 February. *Kapitu* is a period of wet and cold weather with heavy rainfall and frequent floods; even birds have trouble finding food. Even though this calendar is rarely used, the KMB people, especially the small group of fishermen in RT 20, still use their knowledge of the seasonal floods. They noted that in *kapitu*, the west wind brings the high tide that might cause the flood.

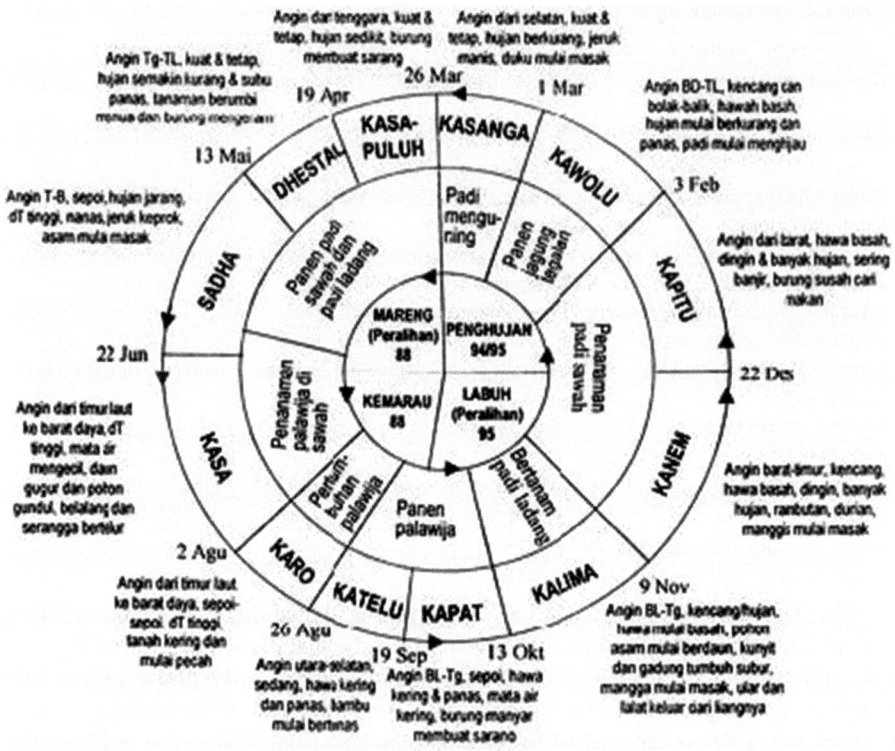

Fig. 4.13 Javanese calendar related to rain (Source: Narso 2012)

Oberkicher and Hornidge (2011) argued that the *lifeworld* provides a stock of knowledge through generating and interpreting individual or collective experiences. The *lifeworld* of flood-affected people is a domain of understanding the decision to live in a vulnerable area. This situational knowledge thus becomes the reference source of KMB's people in assessing their vulnerability to floods. This analysis of the *lifeworld* focuses on the decision of the KMB people to reside in a flood zone that is not equipped to withstand floods.

4.4.1 Shifting Meanings of Floods: From Opponent to "Friend"

The knowledge of flooding is underpinned by the sciences of ecology and hydrology. Floods are caused by the overflow of water from water bodies and/or heavy rainfall (AMS 2012). Thus, the water capacity of a region is important. Many scholars suggested that quantitative data, numerical climate modeling, and flood area mapping are needed to count the numbers of the affected people and to plot the

4.4 The *Lifeworld* of the People of Kampung Muara Baru

direction of the flood flow. Therefore, the flood is categorized as a hydrometeorological disaster that is influenced by climate change, heavy rains, rising sea levels, and high tides. Choi and Fisher (2003) argued that climate change has exacerbated the intensity of floods through increased mean annual and extreme precipitation. The World Bank (2010) reported that annual tidal flooding in the Jakarta coastal area was caused by increased sea levels and high tides, which were worsened by rapid land subsidence.

Based on this conception, flood-prone areas, including KMB, were perceived as a hazardous space to be regulated by zoning rules. The way in which the affected people interpret the meaning of floods should be considered in zoning or other flood management regulations and policies. In KMB, floods are not merely understood as the enemy, but as "a new informal livelihood," "an ordinary thing," and "a playground."

Amir, 21 years old, is unemployed and has less than a high school education. For him, the January flood has become a source of income. He and five of his friends transport rich people from the neighboring settlement in a hand-made plastic lorry that they had found by accident. They earned IDR 950,000[13] on the first day of the flood and an average of IDR 500 thousand[14] per day afterward for the duration of the flood. This amount was more than his usual daily income (Field note, January 2013, translated by the author). I heard another interesting story from "A" who was 23 years old, unemployed, and a resident of RT 19. He said that in the 2012 flood, he prevented the water gate officer from turning on the pump. He even destroyed the water pump to prolong the flood. He did that in the hope of obtaining logistical support for his neighborhood from the government and neighboring private companies (Field note, January 2013, translated by the author). In contrast, Danang, 38 years old, a toy seller who lived at RT 8, benefited from the flood by collecting plastic bottles. Redeeming the bottles replaced the income that he lost from not being able to sell toys during the flood. He used an arched-end stick, net, and rice bag to pick up the plastic. A few other people did the same (Field note, January 2013, translated by the author).

Another story interprets of the flood as an ordinary random event. The women of RT 7 that I interviewed described the flood as something that was both good and bad. "[Flood] is not a peculiar [thing], it is a common that annually visit [her house], but unfortunately never told when came…[laughing]" (Rofadilah, October 2012). "Flood made me tired… But what should we do? For me, the importance is [we must] emotionally prepared… accept the reality…[tiredness] was lost in process of time" (Soliha, October 2012). "Flood could not be guessed, sometimes we thought it's low, in fact it was high, and vice versa…[flood] just likes human beings" (Saparia, October 2012).

The last surprising story shows the meaning of floods for children between the ages of 8 and 14. They enjoyed the flood because they could swim. Besides playing in the water, they caught small fish, snails, and tadpoles. The flood is like a treasure

---

[13] IDR 950,000 = 63.3 Euro (1 euro = IDR 15,000).
[14] IDR 500,000 = 33.3 Euro.

hunt for them; they catch rare types of fish and can share stories with each other. "[We were] happy [when flood time], school is ordinarily off day and we can play until we are satisfied...sometimes we meet many outside people...It was crowded, but many foods were available..." (Irfan, 10 years old, February 2012).

These three stories demonstrate that the meaning of flood is beyond that of a natural disaster. The KMB people viewed the flood as a multicausal phenomenon that continuously reached their operational zone, as something that must be accepted as a normal part of life, and as a necessary reality that was either good or bad. Considering flood is a friend does not mean that they were happy with the big flood, but it is a metaphor representing that the floods sometimes contribute good and bad things at the same time, just like a friend does. It depends on the perceptions of floods and the attitudes toward them. The long experiences of the KMB people have shifted the meaning of flood to be neutral in perceiving this flood reality.

4.4.2 Self-Sensing: A Different Lens in Perceiving Floods

The knowledge of floods has been obtained mostly from quantitative research that is underpinned by ecology and hydrological engineering. Such studies focused on the quantification of data and numerical modeling to produce flood maps and count the number of flood victims. Vulnerability was then defined. However, a distinctive approach is needed to understand people's vulnerability to floods. The *lifeworld* analysis is conducted to understand the perceptions that grow from the lived experiences of flood-affected people and how their subjective experiences are shared. Our analysis revealed different perceptions that resulted from viewing the flood area and vulnerability itself.

First, flood-affected people and the government of DKI Jakarta perceive floods differently. The *lifeworld* of the KMB people is structured by being surrounded by water, the frequency of floods, and the friendships among neighbors. They intentionally choose to live in KMB because of their dependence on the sea harbor for their livelihood. They fully understand and accept the risk of flooding because they have experienced either regular tidal floods or big floods. The government, in contrast, defines KMB as floodplain zone and as an unsafe place to live. According to *Pak Izhar,* the Secretary of the Spatial Planning Agency of Provincial Government of DKI Jakarta, the government has never issued building permits for KMB, especially in the lakeside and coastal side area, which are flood plains. The government has tried in vain to relocate these residents to affordable flats (Interview, 30 January 2013). Hence, the perceptions of the flood-affected people, which emerged from and are constituted by their *lifeworld,* are inconsistent with scientific and technical knowledge. It thus brings attention to the importance of practical knowledge that is exercised over time and of locally situated knowledge, as suggested by Antweiler (2004), to disclose perceptions of flood-related vulnerability.

Second, the analysis found that KMB residents believed that flooding is a natural phenomenon that cannot be changed or controlled. The words *"sudah dari sononya"* mean that it is already given, and *"musibah"* means disaster. Scientists insist that they can limit the amount of rainfall through weather modification technology. The head of *Badan Penelitian dan Pengkajian Teknologi* (BPPT), Jakarta, claimed that the volume or rainfall in Jakarta area in 2013 was decreased by about 20–50% by salting rainclouds (Aziza 2014).

Third, there are different ways of understanding the flood pattern. Based on the *Hijriah* calendar, the KMB people perceive that tidal floods are caused by high spring tides precisely when the full moon comes. Meanwhile, natural scientists have found that tidal floods occur in a periodic cycle of 16.4 months (Hildaliyani 2011). The people of KMB perceived that the heavy rains occurred in *"-ber"* and *"-ri"* months and *"kapitu"* as the time of occurrence based on the Javanese calendar. Scientists developed historical and projection analyses of wet spells as indicators of heavy rains. Therefore, the *Badan Meteorologi dan Geofisika* (BMKG) or Indonesian Meteorology Office has not confirmed the 5-year flood. The head of BMKG argued that floods are caused by high precipitation, which is influenced by evaporation, the Pacific Ocean, and the Indian Ocean (Redaksi-Sains 2011). He denied that the ocean phenomenon happens every 5 years.

However, there is a similar perception of the causes of floods: multifactors (nonlinear) and uncertainty. The KMB people believe that the floods are not caused by the regional factors alone, such as the water flow from Puncak, Bogor, through the Ciliwung River but also by the land subsidence and poor local flood infrastructure, such as clogged drains, collapsing sea dykes, and irregular housing patterns. Similarly, technical experts hold that a combination of global and local factors explains the Jakarta flood. Thus, it is a nonlinear problem. Scientists and the KMB people understand the unpredictability of floods in different ways, which is the reason that preparation for floods is not always optimal. The KMB people do not believe that the day of floods can be predicted; however, scientists explain that it can be predicted. The nonlinearity and uncertainty of flooding perceived by the KMB people and the government support the view of floods as "a systemic global environmental change" (O'Brien 2013, 75).

KMB people also perceive the linkage between flood and climate change, but they equate "nature" with "climate," which they believe is also changing. They observe that it is warmer now than when they were children and that they can no longer tell when it is going to rain. However, numerous studies have defined floods as natural disasters and linked them to climate change, especially to heavy rains, rising sea levels, and high tides. Several scholars argued that climate change has exacerbated the intensity of floods. Climate change is shown by increasing mean annual and extreme precipitation (Choi and Fisher 2003). In coastal areas, rising sea levels and tidal floods are the indicators. The World Bank (2010) found that the annual tidal floods in Jakarta are caused by rising sea levels and high tides, both of which are worsened by land subsidence.

Those self-experiences have developed a such kind of sensory tool for individuals to get cross-cultural meanings of flood. The different meanings that generated by self-sensing and the measuring tools should not be contested, but need to be converted in the same language of flood-related vulnerability. The self-sensing of the KMB people can be used to help in measuring the level of vulnerability.

4.5 KMB People's Perception of Flood-Related Vulnerability

The insights of flood-affected people can be used to define flood-related vulnerability. The IPCC (2012) used exposure, sensitivity, and adaptive capacity to define vulnerability. As discussed in Chap. 2, I use the interrelation of those factors as the framework for measuring vulnerability from the perspective of flood-affected people. I assume that the floods will probably keep changing the impressions and the perceptions of the KMB people. Having discussed the structure of the *lifeworld* of the KMB people in the previous section, in the following sections, I assess the three factors of flood-related vulnerability that were elaborated by the IPCC: exposure, sensitivity, and adaptive capacity.

4.5.1 Perceived Flood Exposure

Based on interviews with the heads of RTs, I identified how they perceived floods in their area. My assistant and I interviewed 22 heads of RTs and asked them to describe the biggest floods in each year from 1992 to 2012. I brought a map, conducted transect walking, and asked them to delineate the inundated area and describe their own impressions of the floods. The interviewees described the exposure in terms of frequency (occurrences of big floods in a year), magnitude (size of inundated areas), and intensity (flood level).

The leaders of the neighborhood associations claimed that floods had become less frequent since 2007 because the construction of a 1.5-m seawall (measured from the land surface) has prevented tidal floods. They claimed that since 2007, there have been no more than five floods a year. Previously, they had had an average of 7 to 8 floods per year. However, inundation still occurred when there were heavy rains, but they did not categorize it as a flood. "If [the rain lasted] 5–6 hours, the road was inundated because the drainage was clogged…" (Nanang, interview with the author, 7 December 2012). Similarly, another interviewee, Pak Nasirudin (RT 20) said: "here, [if] a half day is rainy, the water [is raised] ankle-deep. But it disappeared quickly" (Interview, 8 December 2012).

They based the intensity of floods on the physical dimensions of an adult: over the head, chest-high, waist-high, knee-high, and ankle-deep. They always measured the depth of the water on the road in front of their houses. The findings showed that the most intense floods were in 2002, 2007, and 2013 (January). The volume of the water was relative, but the 2013 flood took about 6 days to subside, which was longer than the 2007 and 2002 floods, which only took 4 days. When asked about flood distribution, the KMB leaders claimed that most of their area was

4.5 KMB People's Perception of Flood-Related Vulnerability

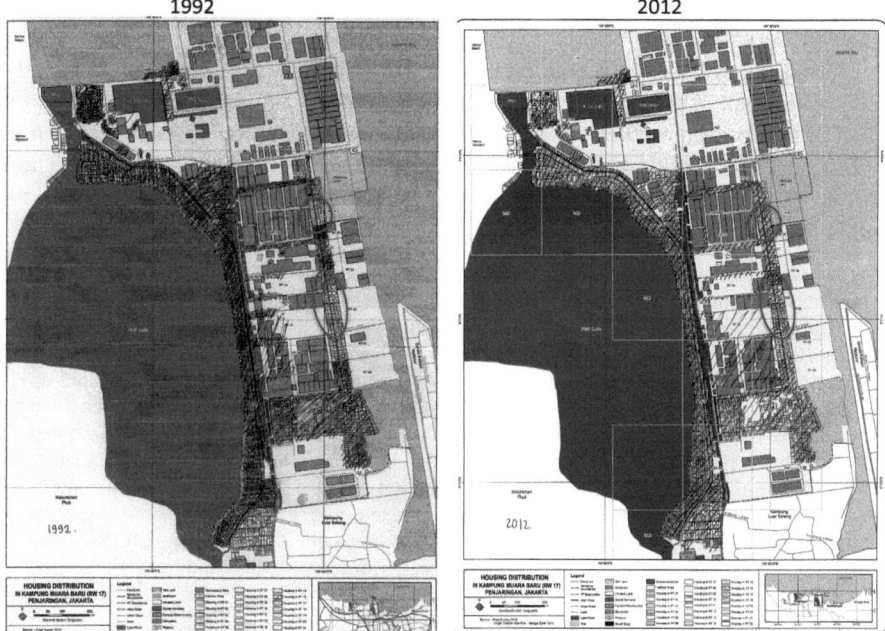

Fig. 4.14 Flood map in 1992 and 2012 as perceived by the neighborhood leaders (Source: the author, based on the interviews with 24 head of RTs and transect walks (2012))

badly inundated, except RT 2 and RT 3, which had the shallowest water. The heads of RT 16 to RT 20, which are located next to the sea and Pluit Lake, thought their areas flooded more frequently than other RTs on the right side of *Muara Baru* Street. The community leaders claimed that between 1992 and 2012, there were more floods, especially in RTs 14, 12, 06, and 03 (see Fig. 4.14).

My presence in KMB during the flood of January 2013 allowed me to conduct participant observations to confirm the perceptions of the KMB people. After 2 days of heavy rain that began on Tuesday afternoon (15 January 2013), KMB flooded on Thursday (17 January 2013) after the Pluit embankment was perforated and the water pump was broken. I visited the flooded location on Friday (18 January 2013) around 7:00 a.m. The floodwater was still high, and no mass logistical support had arrived. People were moving from the KMB area to the *kelurahan* office, the main road, and the mosque to find food, salvage their belongings, or simply rest. According to Konedi, Secretary of RW 17, there had been no warning, and no one had expected the flood to be so powerful. He predicted that the waters would be only knee-deep and then recede. However, on Wednesday night, around 9:00 p.m., he learned that Pluit Lake was flooded. The people in RTs 18, 19, and 20 had moved to the Harbor office and public apartments, which are on higher ground and have multilevel storage. By six o'clock the next morning, the level was up to the neck and had covered the entire KMB. All RTs were evacuated on Thursday morning (17 January

2013) to the main road, mosque, police station, and subdistrict office, where the water was not as deep.

During the evacuation, I noticed that the KMB people were behaving normally. They walked in groups, carried their valuables, and chatted with their neighbors. Most of the men sat on their roofs, but the children and women evacuated. The men went to the second floor of mosques and schools and waited for assistance. There was no panic; everyone seemed to know where to go and what to do. I wrote the following in my field notes:

> I just waded through the flood, I saw many people walking in the opposite direction, bringing their belongings while occasionally talking to each other and saying hello to other friends that they met in the street. [Unconsciously] I heard someone call my name, and he was *Pak*. A, the head of RT 1, he smiled and shook my hand and asked why I was here. There was no sadness or panic in his face. He wore a long-sleeved white shirt and rolled-up black trousers, as usual. It looked like he didn't change his clothes last night when the flood came. We talked for about 15 minutes. He told their evacuation stories, he emphasized that he and his neighborhood had known since midnight, but they decided to evacuate because the water was getting higher, and they heard the Pluit dam had broken, so early in the morning they left when the water was at knee level. He told that it was just the same five years ago in the [2007] flood.... He also explained what he would do if flooding still inundated their place. He gave the impression that he had experienced with this condition. [Field note, 17 January 2013]

For people who had been in KMB for more than 5 years, floods are not major disasters. Long-time residents know that they live below sea level and in a region that is easily flooded. Tidal floods, rainy floods, and floods transferred from Bogor are common. The tidal flood event is part of the Muslim calendar, the rainy flood event is part of the Javanese calendar, and the transferred flood is always discussed in the mass media and local conversation. These perceptions constitute the realm of the KMB people. However, the perception of themselves as vulnerable and as victims was belied by their responses to and behaviors during the flood that I experienced.

In conclusion, based on their place-based memory, the KMB people know that they live in a flood-prone area. They know that the place is easily inundated, becomes a disposal site, and stays wet. They admit that the danger is not the frequency of flooding but the government's late responses to it. The January 2013 flood confirmed that they tend to look for logistical help rather than worry about their inundated houses. Konedy, the Secretary of KMB stated, "We are mainly vulnerable to the lateness of flood aid, not the magnitude of floods" (Konedi, interview, 20 January 2013). In addition, "the severity is the worst, but people complain about the lateness of logistical support, not their severe condition caused by inundated houses or loss and damage due to floods" (Akuntono 2013).

4.5.2 Perceived Flood Sensitivity

This study used a people-centered approach to identify the sensitive people in KMB. People-based sensitivity means a given condition of individuals or groups, such as age, gender, health, and origins. This condition influences them in coping to or adapting to the floods. The KMB people confirmed that the elderly and children under 6 years old, people who are sick and/or disabled, and new residents are sensitive to floods.

Ibu Rofadilah (RT 7) said that the 2007 flood killed a 6-year-old in her neighborhood. The child was playing in the water in front of his house. Because the alley was too narrow and the [facade] buildings were irregular, there was no a safe hiding space when the floodwater flew fast. The water threw him against the pillars of the house, and he drowned. (Interview, 07 February 2013). A toddler died in the 2013 flood. As told by Ninu (28) in the media online, the victim was one of 11 flood-affected infants who were evacuated and died of fever and diarrhea on the fifth day of the flood (Winarno 2013) Both stories were shared by the KMB people in the shelters and later in their neighborhood. It taught them to act preventively with their children because they understood that they were sensitive to floods that lasted for more than 5 days.

An elderly man with no family died because he was too weak to escape the flood. Gustara, the head of RW 17, said that two elderly people died in the 2013 flood. One was Tjasti, 74 years old, a resident of RT 08, and the other was Salim, 60 years old, a resident of RT 10. Both had been ill, but they refused to be evacuated. However, there were not enough food and medicine in their home because they had probably not expected the flood to last 5 or 6 days. Gustara said, "Elderly people are sensitive to floods because they are dependent people [who] should be accompanied by their family members or neighbors" (Interview with the author, 23 February 2013).

KMB people understand that the sick are sensitive to floods. They are easily infected by the unsanitary conditions caused by the floods. In addition to the 5 or 6 days of flooding, the garbage and pollution from other parts of Jakarta contribute to the danger. The water is black during a flood. Long floods are followed by blackouts, so the nights are cold and mosquitoes are everywhere. Agus said, "If [the flood lasts] 2–3 days, [we are] still fine, can survive, no bath, no changing clothes, limited food and water, but if it's more than that, just like last January before, the healthy person can be sick, moreover the sick people" (Agus, interview, 24 February 2013).

People with disabilities are also sensitive to the floods because their mobility is limited. Asep, the blind masseur in RT 16, depends on his phone to call his friend who lives in RT 20 if a flood comes. He needs to be guided in packing his important belongings, dry food, water bottle (aqua), clothes, radio, and his stock of massage potions. During a small flood, he navigates by hitting his stick against the pillars of houses and listening to the voices of *bajaj* or truck sounds to find the main road (*Muara Baru* Street) (Interview, 7 February 2013).

The final group of sensitive people mentioned in the interviews and the two group discussions held in RT 15 and RT 20 are new residents, house owners, and land-based people. The new residents are individuals who have been in KMB for less than a year and who have never experienced a big flood. These people are very sensitive because they tend to panic. They usually have limited social relationships with neighborhood leaders, who often have better access to and information about logistical support. Newcomers are likely to become confused when a flood strikes. As a result, they usually lose their belongings on the way to the shelters because they are too heavy and already wet. The house owners spend money to clean and improve their property, but tenants or renters leave their dwellings. Therefore, the losses incurred and damage suffered by new residents and house owners are usually greater than those incurred by the renters. People who work on the water, such as fishermen or boatmen (*manusia perahu*), are not as bothered by floods. In contrast, the land-based people, mostly the second-generation KMB people who work downtown, are clumsy and heavy-handed in responding to the tidal floods.

Thus, the sensitive people know that their settlement is irregular, which causes problems in getting logistical assistance from the outside into the neighborhood. They also know that the elevation of the houses, which is under the level of the main road, means that their area is easily inundated. However, they still perceive that the sensitive condition of their living place is not their greatest worry. It is more important for sensitive people to know the neighborhood environment and have the ability to contact friends, not panic, and seek assistance.

4.5.3 Flood-Related Adaptive Capacity

Adaptive capacity can reduce both exposure and sensitivity. However, the human capacity to adapt to flood situations is a key element in building a resilient community and adapting housing and settlement to flood situations. Therefore, the KMB people must define their own adaptive capacity.

In the in-depth interviews with several people who have lived there for more than 10 years and group discussions held in RT 7 and RT 15, the word "experience" was often expressed. Experience is the best teacher. Residents use not only their own experiences in adapting to a flood but also those of others. These experiences include the daily activities during floods (responses), post-flood (recovery), and the next flood (anticipatory). The responses that they have during floods are eating and drinking (cooking or eating), getting their children to school, earning a livelihood, and sleeping. To do these daily routine activities, individuals or groups need practice. A past action is used to create a better future action. Both the holistic understanding of the KMB neighborhood and the strong relationships with neighbors help the residents to adapt. Therefore, the longer they live in KMB, the more familiar they become with the safest places, the shelters, the fastest evacuation route, and the best survival strategies.

4.5 KMB People's Perception of Flood-Related Vulnerability

Experiences in relation to recovery activities pertain to cleaning, repairing, and replacing lost or missing belongings. Less experienced people spend too much on repairs and do not know how to find inexpensive replacements. Fendi (RT 16) stated, "Generally people clean their houses after the water gone, which left a lot of dirt, but it is wrong because it must be not enough water [to clean]… the good [time to clean] is when the water was still… at least reach ankle high" (Interview, 28 January 2013). Ridwan (RT 20) said that he always puts off repairing damaged parts of his house because construction materials are part of the assistance that he receives. When I first met him, he thought that I was a donor or a donor's representative because I asked him what had been broken. He said that aid was available, but people have to know how to get it. New residents do not always know this. According to Fendi (RT 16), Ghufron (RT 17), Sahrul (RT 21), and Endang (RT 20), at the end of January 2013, many pieces of furniture and materials were floating around the neighborhood. Teenagers could earn money by collecting plastic bottles and glasses, making carts and renting them to deliver people or motorcycles, or even selling the materials in the flea market during flood sessions. Therefore, experience is essential for adaptation.

Furthermore, the ability to anticipate floods is related to keeping houses as clean as possible, creating a storage place, preparing food, obtaining logistical support, and improving neighborhood facilities. The experienced people compare the height of the water to their height of the floor. Some people can afford to raise their houses. Those who cannot raise the floor build a dyke in front of their door or install the temporary storage on the walls. They also stockpile gallons of water gallon and *indomie* (instant noodles) in anticipation of the next floods. Dorkas and his wife Santi (RT 15) learned from the 5-day flood in 2007 that they could both survive on one gallon of water and half a carton of instant noodles (not boiled). Moreover, Konedy, the Secretary of KMB, always evaluated the latest flood with the heads of the RTs, seeing which facilities had and had not been used and which facilities were the most in demand but not available and seeing the performance of shelters and the logistic mechanism. They did not hold a public forum, but met in small groups in the secretariat. They never discussed building facilities against the floods or try to reduce their magnitude, intensity, or frequency because they knew that they did not have the capabilities and competencies. The community leaders were more concerned about surviving the floods and minimizing their losses.

For poor residents or renters, activist or solitaire, or workers or the unemployed, the experience can reduce vulnerability and increase adaptation. The extent to which this is possible depends on how the boundaries of experiences are constituted in their life. Some people limit their zone of operation to their family and house, but others extend their zones to the neighborhood. The story of Agus, head of RT 15 (see Fig. 4.15), shows that people can transcend their *lifeworld* boundary when it is necessary to work for the neighborhood.

> Agus, 28 years old, born in KMB, living with his five family members in 21-square meter wooden-house, attached to the sea-dykes. He does not have a job, but he rents three rooms on the second floor. He has experienced floods both in his childhood and on the present. He senses flood just like ordinary events, not a big deal for him. He remembered when he was a child; he always played in the floodwater in the rainy day with his friend. It was a happy day, he said. He sees from time to time, the flood is just the same, but it takes a little bit longer to be dry. Now, he has responsibility to keep his family safe since his father went back to their village. When January flood came, he told his struggling in finding the food because his stock was diminishing. With his two male friends, he created floating cart from plastic fiber, which found in the neighborhood and went to main road to find the logistic truck. He found army-truck was stopped, but not distributed the food. He afraid to ask, but because of hunger, he just forced the army to give them more foods. He just jumped in the truck and threw the noodle packs to his colleagues. He succeeded to bring the instant noodle, aqua, bread, etc. to his house where his family and several neighbors stay on the roof.

Fig. 4.15 Adaptive capacity of Agus, the head of RT 15

4.5.4 Adaptive Capacity of Agus, the Head of RT 15

Although the houses and livelihood of the poor are fragile, the intensive lived experience is their main asset. The experience of living through a flood gives them a place-based identity and established social relationships, which improves their survival capacity during a flood and their adaptability in the KMB area. Some residents of KMB stay in place and adapt to the conditions. They are constantly adapting, such as by raising their homes or changing their habits in response to the stress and shock caused by the flood. However, other people, especially tenants or renters, move temporarily; some leave KMB and find another livelihood.

The highly lived experiences of the KMB people can be gained only if they are socially active in their neighborhood. KMB has an informal union for women (*arisan*), Muslim solidarity (*pengajian/majelis taklim*), and neighborhood talks that allow for sharing among the members and neighborhood improvements, such as clearing small drains before and after floods. The sharing is based on the Indonesian spirit of *gotong royong*. The entire community contributes resources and participates in a project or activity that benefits the community or specific members. It supports Jelinek's argument that the *kampung* has the cultural value of *gotong royong*, which distinguishes it from urban settlement.

Gotong royong is a traditional spirit rooted in Indonesia's cultural values. It is "a mutual social action institutionalized in a community, rather than one born from the full voluntary will of villagers for mutual support" (Kobayashi 2007, 9). Kobayashi argued that *gotong royong* is a constructed tradition that depends on the community. As a community of multiethnic migrants, the KMB people maintain the spirit of *gotong royong* while building their adaptive capacity. They institutionalize *gotong royong* though *kerja bakti* in each RT. *Kerja bakti* can take the form of cleaning or repairing drains and other neighborhood facilities. KMB people sometimes conduct *kerja bakti* for public projects, such as mosque construction, or communal events, such as marriages and deaths. Therefore, the KMB people use *kerja bakti* to update information, informally discuss issues, and maintain their solidarity to increase their adaptive capacity.

4.6 The Continuum of "The Vulnerable" and "The Adapters"

In the previous chapter, I mentioned three concepts derived from the vulnerability framework: exposed people, sensitive people, and adaptive people. I placed these concepts along a continuum of vulnerability and adaptation. Adapters have the greatest adaptive capacity because they use all resources at their disposal and can transform their vulnerability into adaptability. The *lifeworld* is the domain of increasing adaptive capacity (see Fig. 4.16).

Exposed people are similar to flood victims. The level of flood exposure is derived from the comparison between past and present floods, personal involvement, and the amount of loss and damage. The number of floods experienced is thus very important to understand a person's degree of exposure. Based on the flood experiences of KMB, I proposed a self-comparison method to define flood-related vulnerability. To conduct the self-comparison, I needed to know the spatial, temporal, and social arrangements of individuals in KMB. The place-based memory of the situation of floods was an essential source of their historical and comparative flood exposure. This memory revealed their perceptions of flood exposure. The depth of the perception depends on how long they had stayed in KMB and the closeness of their relationships with old residents. The people who had lived there for decades had the strongest sense of being surrounded by water. They understood that their world is bounded by water and temporarily structured by a series of inundations. In contrast, people with limited experience had a weaker sense of exposure and less access to information about floods.

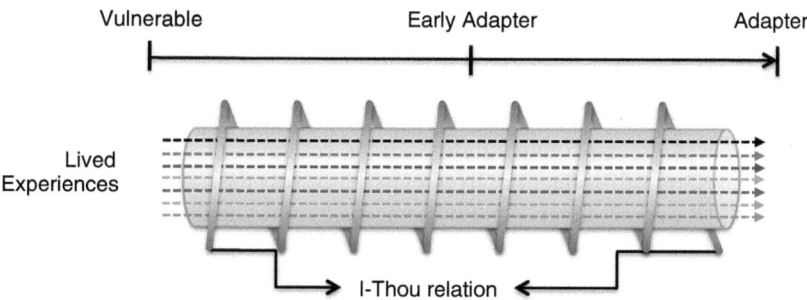

Fig. 4.16 Vulnerability assessment using lifeworld analysis

Sensitive people experience more losses and damage than the adapters experience. In the medical world, people who are allergic to some medicines can be weaker than others are. The assessment of sensitivity is based on their circumstances. The KMB people have similar ideas about sensitivity. People who depend on others are the most sensitive. Their zone of operation is predictable and often is inadequate to cross the flood boundary. The source of the actions is the world of their predecessors. KMB people have identified the sea-based culture and livelihood, land entitlement, age, and health and disabilities as the main factors that can influence sensitivity. KMB people also demonstrated that these factors change if people share their experiences and teach others to compensate for the limitation of their zone of operation in a flood.

The poor need to increase their income, but they do not necessarily depend on their economic capacity to adapt to the flood. They can count on experience. The KMB people use the flood's experiences to adapt, as did Agus and Amir. The escape route, neighborhood cooperation, and how they find shelters and build preventive facilities are examples of the creativities of their adaptive capacity. Because of these practices, floods are not considered extraordinary. Therefore, adaptive people know how to live through floods. As previously discussed, local innovation comes from their limited capacity through trial and error to protect their houses and neighborhoods from the inundation. These experiences make them self-reliant in adapting as individuals or as a group.

Therefore, flood-related vulnerability changes periodically. It could be reduced if based on experience, the creativity and innovation are able to adapt to the exposure and sensitivity. However, vulnerability also can be higher if the poor have little or no experience in coping with the floods. Adaptive capacity decreases vulnerability. It relies on how the intersubjectively shared world of experienced people is constituted by the poor. Through *kerja bakti*, the KMB people can increase their adaptive capacity because *it* is conducive to sharing. It creates a stock of knowledge that can be applied and reapplied by the KMB people to increase their adaptive capacity. Therefore, I argue that the poor are not vulnerable if they can use their stock of knowledge to improve their own and their community's adaptive capacity. In fact, they can be adapters if the enhancement of their adaptive capacity is able to surpass the degree of exposure to flood and the value of their sensitivity. They can be more vulnerable if they fail to capitalize on the knowledge sourced from their social world to manage their exposure and sensitivity.

Based on the in-depth investigation on the *lifeworld* of flood-affected people and the analysis of the *lifeworld* perspectives, I found that not all the poor were vulnerable and that living in a vulnerable area did not make a person vulnerable. There were different degrees of vulnerability among residents of the same building. In addition, based on self-reflection, lived experience was the central factor that influenced all KMB people, tenants and owners, males and females, the young and old, and the socially connected and isolated. Life experience distinguished the adapters by their preparedness, knowledge of safe places, ability to survive on a roof and clean up afterward, and capacity to be mobile. KMB people defined vulnerable people as those who could not do these things. Moreover, the *lifeworld* of the KMB

4.6 The Continuum of "The Vulnerable" and "The Adapters"

Table 4.3 Lifeworld analysis of flood-related vulnerability

	Vulnerable	Early adapter	Adapter
Exposed people	Being threatened (pre-flood), gets panicked when evacuated (during flood), decides to move to other places (post-flood)	Being casual	Being casual
		Reacts normally	Still does his/her habits
		Willing to move to another affordable place	Remains and does not want to move out
Sensitive people	Dependent on the others	Independent to some extent, can take care of his/her personal matters	Independent and can assist the others
Adaptive capacity of people	Inexperience	Few experiences	Experiences not only flood but also knows the way out
	No idea about the detail situation of his/her surrounding,	Survival responses during the flood	Recovers quickly
	No creativity and innovation	Following the others who has creativity	Generates income during flood in a creative and innovative way

Source: the author

people had constructed a meaning of flooding that is completely different from that of the natural scientists. KMB people see floods as common and frequent events. They have no idea when they will strike or how bad they will be. In contrast, natural scientists describe floods as periodic hydrometeorological disasters. There is a 5-year cycle of big floods and a 16.4-month cycle of tidal floods. The causal factors are heavy rainfall and high spring tides that the ground cannot absorb, which are exacerbated by improper land use, land subsidence, and poor drainage. The different perceptions influenced the identification of vulnerable people and defined both their vulnerability and the adaptation pathway that needs to be followed.

As stated in the previous chapter, a people-centered approach can be used to assess vulnerability, in addition to the previous eco-place and socio-spatial approaches. In this study, I argue that *lifeworld* analysis can offer an alternative way to identify vulnerable people and places. It has the advantage of providing individuals' perceptions of not only the state of vulnerability but also the role of adaptive capacity. This analysis does not use the usual parameters of social vulnerability (e.g., health, education, economic condition) as in the quantitative approach, but it reveals the factors that have meaning for the flood-affected people. Table 4.3 shows that the adaptive capacity of vulnerable people could be optimized by the number of flood experiences and the amount of reciprocality achieved through their own mechanisms of social relations. It transforms them gradually from the state of being vulnerable to being adapters.

The vulnerability assessment is a central component of an adaptation. It is important in gathering information on "what to adapt to and how to adapt" (Füssel and

Klein 2006). The lifeworld analysis contributes to elaborating the state of individuals in adapting to the changing phenomena that they have experienced. This internal assessment emphasizes the meaning that people assign to their experiences. Because those meanings are shared, the vulnerable state of the community is the aggregate of individual vulnerability.

The global awareness of the utility of employing local knowledge in the vulnerability assessment emerged through the SREX report of IPCC in 2012 and the Global Platform Consultations report on Post-2015 Framework for Disaster Risk Reduction (HFA2) in 2013. UNISDR recommended that the development agenda for building community resilience should focus on reducing vulnerability and strengthening capacity. The recent discourse relies on socio-spatial and eco-place approaches. Therefore, research is needed to assess the vulnerability and the adaptive capacity from a people-centered perspective. This study offers a *lifeworld* analysis that assesses vulnerability. This case study of the urban poor who live in a flood-prone area found that the human dimension is essential in measuring a community's vulnerability. Further studies that examine different communities and other effects of climate change effects and types of disaster are needed to obtain a holistic understanding of vulnerability assessment.

Chapter 5
Locally Embedded Adaptation Planning

Our people seldom read...and write even less.

Konedi, secretary of KMB

Abstract The chapter elaborates what locally embedded adaptation planning (LEAP) is, its meaning, and how it is institutionalized in *kampung*. The discussion focuses on the lived experiences of individuals and groups in planning their housing management, evacuation and shelter strategy, flood infrastructure provision, and income generation. Following that, this chapter explains how their planning activities become habitualized actions and how they reciprocally typify their planning. This chapter discusses the presence of informal planners who transmit planning knowledge through several nonformal events in the community. The unwritten communication has played a significant part in producing the planning knowledge.

Keywords Local • Knowledge • Kampung • Institutionalization

The chapter elaborates what the meaning of locally embedded adaptation planning (LEAP) is and how it is institutionalized in *kampung*. The discussion focuses on the lived experiences of individuals and groups in planning their housing management, evacuation and shelter strategy, flood infrastructure provision, and income generation. Following that, this chapter explains how their planning activities become habitualized actions and how they reciprocally typify their planning. This chapter discusses the presence of informal planners who transmit planning knowledge through several nonformal events in the community. The unwritten communication has played a significant part in producing the planning knowledge.

© Springer Nature Singapore Pte Ltd. 2018
H.A. Simarmata, *Phenomenology in Adaptation Planning*,
DOI 10.1007/978-981-10-5496-9_5

5.1 Reflection on Experiences and Learning

In order to know whether the practices of KMB people involve a planning process or not, I discuss my research findings by combining Schütz's lifeworld theory and Schön's reflective practices. As described in the conception framework, the *lifeworld* is a domain of the reflective practices at work in the community. Departing from Schön's argument that planning process needs to "shift from rationality to reflective-in-action" (Schön 1983, 21), increasing the capacity of planners through enhancing the ability to reflect in action (Fischler 2012), and educating the planners by a method of "learning through personal experience" (Dewey 1998, 9), I show how the KMB people apply these concepts in relation to adaptation planning.

Schön's idea of reflective practice is rooted in examining the relation between the thought and action of anyone who plans something for anything. It is the kind of thinking that shapes our actions before, during, and after the action. The quality of an action then depends on the thinking that we can do before and in the process of the action. Therefore, in this research, the main discussion of reflective practice concerns the ability of KMB people to use their own and others' knowledge to improve their adaptation actions in real time. Furthermore, I discuss the production of knowledge of KMB people based on the "triple-loop learning process" inspired by Aagrys and Schön in 1978 (see Fig. 5.1).

As argued by Pahl-Wostl (2009), triple-loop learning is an exploratory process at different levels of intensity and scope. She described three levels of learning that take place in analyzing adaptive capacity:

> Single-loop learning refers to an incremental improvement of action strategies without questioning the underlying assumptions. Double-loop learning refers to a revisiting of assumptions (e.g. about cause-effect relationships) within a value-normative framework. In triple-loop learning, one starts to reconsider underlying values and beliefs, worldviews, if assumptions within a worldview do not hold anymore. (Pahl-Wostl 2009, 359)

People learn from floods. In the first phase, learning creates new knowledge that is likely to confirm the practices of others. It shifts the meaning of what the previous

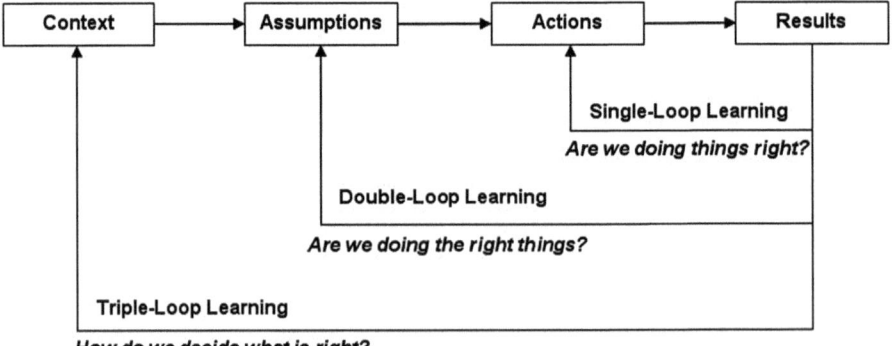

Fig. 5.1 Model of triple-loop learning (Source: thorsten.org)

practices have achieved and transforms the paradigm of the adaptation. Therefore, there are two main parts to the following discussion: how reflective practices reflect the process of KMB people in adaptation planning and the type and source of knowledge that drive them to plan and its relation to their *lifeworld*.

The shifted meaning in the realm of the KMB people who adapt to floods plays a significant role in shaping their adaptation pathway. The shifting meaning of floods has transformed their response to floods. They repeatedly adapt. As a result, the flood is no longer perceived as something that needs to be blocked or stopped but should be survived with as little damage and inconvenience as possible. KMB people produce a simple plan based on the reflection on their last experiences and their present and/or future adaptation. It is a plan that provides lessons learned from the series of adaptation actions that are accumulated, articulated, and preserved in memory. The lessons learned strengthen their intuition. This intuitive knowing is the main modality used to increase the adaptive capacity of KMB people. Here, I discuss the planning knowledge used by KMB people in building their adaptation pathways through three planning practices: scenario planning for evacuation and shelters, spatial planning for living space arrangements, and infrastructure planning for neighborhood support.

5.1.1 Scenario Planning for Evacuation

Scenario planning is "a creative and shared process" that allows people to reflect on "what seems to be happening (analysis), what's really happening (interpretation), and what might happen (prospection)" (Conway 2003, 20–31). Unlike prediction, scenario planning uses a foresight method that tells about what might happen. According to Conway (2007, 7–8), "Everyone has an innate foresight capacity to think about the future." In organization learning, foresight used "to assess the implication of present actions, to detect and avoid problems before they occur, considers the present implication of possible future event, and to envisage aspects of desired futures" (Conway 2003, 16).

In planning evacuation routes, KMB people have engaged in scenario planning because they used foresight method to sense the uncertainty of future floods. Based on their flood experiences, they discuss the consequences of the previous evacuation process and the obstacles to staying on rooftops, and they envisage the next steps for the future. As explained in Chap. 5, negative experiences have taught them to be more careful in the evacuation. Therefore, they prefer to take shelters on roofs. If the flood reaches the roof, they will evacuate. Reflecting on what they had done before, they keep anything (e.g., floating plastic carts) that was available to use in the next flood. They have no resources to escape, but when the worst case occurred, logistical supports, such as lifeboats and life jackets, were available (Field note, January 2013).

In arranging the evacuation plan, KMB people develop a scenario that reflects their past experiences with floods. In my group discussion with RTs 7, 15, and 20,

Table 5.1 Scenario planning for flood evacuation

Flood scenario	Plan	Logistic need	Evacuation tools	Remarks
Knee-size inundation (less than 50 cm)	Stay at home; manage house by own	Modest food for family	N.A.	Focus on the protection of their electric equipment at house
Chest size (1 m)	Transfer to the two-story house of neighbor	Modest food, water, and cooking equipment	Wood stairs and flashlight	If the logistics run out, borrow from neighbor (*warung*)
Adult-body size (1.5 m)	Gather in the roof of two-story house	Modest food, water, and cooking equipment	Life jacket and life boat (if any); plastic or PVC cart; drums	If the logistics run out, send young man to get the food outside
				The tools are available in the secretariat of RW
Over the adult-body size (the same as the roof level)	Leave out the house by emergency boat from the government/donor	None, probably hand phone if still alive	Life jacket and life boat (if any)	Waiting for the instruction from the head of RT; evacuating the women, children, and elderly at the first time

Source: the author based on the focus group discussions of RTs 7, 15, and 20

the consensus emerged that if adults could not keep their heads above water, they would leave their houses. Otherwise, they would go to a nearby two-floor house (see Table 5.1). Agus, the head of RT 15, said that if the inundation lasted only 2 or 3 h and the water was not above the knees, the flood would recede within a half day. This happened with tidal floods or during rainfall in the wet season. However, these floods are becoming rare because of the seawall built in 2007. The floods coincide with the full moon (tidal flood) and wet season (rainy flood).

Agus recognized that it was tiring to carry his television to and from the second floor. He realized that it was useless because the flood only reached the table leg and receded quickly. He indicated that his electronic equipment would be safe as long as the waters remained below the knee. He also told me that as the head of RT 15, he has shared his experiential knowledge with his neighbors and at several neighborhood events. The knowledge about protecting equipment was confirmed by the FGD participants in RT 15. I saw heads nod in agreement when I asked what they would do during a flood. Khairuddin, one of Agus' neighbors, added that he used his leg to gauge a flood's intensity. He asserted that water that did not reach the knee was not worrying.

Another example was raised in the discussion with RT 20 about evacuation. Irpan, the head of RT 20, described a neighbor's scenario plan that was similar to Agus' plan. However, because their area is surrounded by the seawall, the first

5.1 Reflection on Experiences and Learning

scenario of floods seldom occurs in their area. He knew the flood behavior in his area. Pak Irpan's only concern was that the land was sinking, which was raising the water to the top of the seawall. He wanted the seawall elevated because it was only 20 cm higher than the water. He warned that if the seawall were not elevated, the flood damage would be similar to that in worst flood of 2002.

> A: How do you assess the flood in your area?
> I: First, I always figure out when flood comes in and out as well. The flooding (period) usually takes a long time for the sea emptying (means receding). Second, we know we have a sea-dike, but if it is broken or overflowed, then we wait until the water recedes, but no need until the water run out because the water recedes slowly. (Irpan, interview, 27 October 2012)

He added that the 2002 flood receded in a month because the Pluit Reservoir and the generator for the water pump were not working. The flood was caused by the combination of the high tide, the flood transferred from Bogor, and the extreme rainfall that did not stop for 24 h. The water from Bogor could not flow to the sea because the sea surface was at the same level as the inundation height. He pointed out that using the water pump was useless.

> A: So you mean that the sea wall was never elevated?
> I: It has been done already, but broken in 2007. The height of water was about 1.5 meters come from this way [he pointed to the front of his house that faced the sea]. I hope that I won't experience it again. But now, it's getting harder. Our position now is two meters below the sea level; therefore, I have warned my people where next to the sea-dyke, if it starts to be cracked, I ask them to report me. [He is silent for a moment.]
> By the way, are there any benefits for the residents here that taken from this research? Or is it just a manuscript? I need your help to inform the government because sometimes university students have interviewed us. If high tide comes, the difference between sea surface and dyke's height is only 20 centimeters. When I started to live here [1990], the elevation was about 0.25 meters under the sea surface; now [2012] it has already reached two meters. (Irpan, interview, 27 June 2012)

Ibu Maskuni, who has lived in RT 7 since 1983, noted some differences from before the 1990s, when KMB was still empty and the houses were made only of plywood, and the present condition, in which factories, warehouses, and houses thrive and the area is dense. She explained that she has two flood houses made of concrete, but she worries because the floods last longer and evacuation is harder. According to her observation, the buildings blocked water flows, and the narrow alley and the housing density increased the difficulties for rescue boats and finding the fastest evacuation route.

However, she described her evacuation plan to me on 4 July 2012:

> A: Have you experienced floods entering your house?
> M: Yes, in 2010, when the dyke was still made by sandbags, it reached chest-level [she demonstrated by putting her hand on her chest] and even a rubber boat entered the neighborhood.
> A: What? How could that happened?
> M: It was possible, because not the big one.
> A: Were you shocked at that time?
> M: Not shocked, but I hesitated whether the water would keep rising or not.
> A: Do you remember when?

M: If I am not wrong, on Isra Miraj day [the Moslem day].
A: In 2010?
M: Yes, in July or August.
A: Daylight or night?
M: Day and night. It lasted 10 days.
A: Did you evacuate?
M: (No), just waited on the top of the roof. I was afraid my belongings would be stolen or swept by flood. So, we just lifted them up.
A: Children too?
M: Yes, of course.
A: Did you take a bath?
M: Sometimes (laughing), to take a bath, we just plunged [into the water].
A: For drinking?
M: We bought Aqua [Aqua is a brand of drinking water that quite famous in Indonesia. It usually used as a generic term].
A: Where and how did you buy it?
M: We bought it by ourselves, in the main road [Muara Baru main street], because only this area was flooded.
A: I mean how do you reach the place? The water was on your chest, right [I pointed to my chest].
M: Yes, we just walked slowly to find food and drinks for our children especially.
A: How do you find the route?
M: We got used to taking detours because it safer from our experiences.
A: So you got wet?
M: Yes, of course we got wet [laughing]. Our clothes, underwear… everything got wet.
A: But you still did it?
M: Yes, it's fine. We admittedly were sick, but we must did it and it's not so difficult. We already knew our way out.
...

In this neighborhood, Pak Kadir, the head of RT 7, described the same evacuation plan. He added that even though one of his neighbors had learned to walk to a safer place, his people had their own evacuation route, which he described in an interview on 23 June 2012.

...
A: As the head of the RT, do you have a rescue or early warning system?
K: No, we don't have it at all. If the flood comes, usually the rubber boat comes.
A: From RW?
K: I don't think that our secretariat has it.
A: So, from where?
K: I usually contact the head of RW, but if it does not work, I send our young men here to go to the main streets to find the army or government people.
A: So, is there any training for emergency responses?
K: Yes, there was, but I did not attend, the representatives of our people that came to the training. It was in Puncak [Bogor] and/organized by Indonesian Red Cross [PMI].
A: In here? In your neighborhood?
K: Yes, once I think. It was under PMI as well. But I forgot the names.
A: Did women here follow the training?
K: Yes, most of us followed the training. We followed the tsunami simulation. How to evacuate, find kentongan [the traditional tool made by wood that used for indicating the clock. It is originally used to monitor the neighborhood security at night], etc.
A: But does it work? People still recognize the sound? Why don't you use a hand phone, maybe?

5.1 Reflection on Experiences and Learning

K: Yes, indeed [laughing]. Everybody uses a hand phone, but it also sometime does not work because when flood comes, there is no electricity. We don't have kentongan as well, but I think people have already known by themselves because the flood is rising slowly. So, we still have time.

A: Did you get another lesson?

K: Yes, we were taught to run to the safety place [laughing]. It was fun, but we never did that.

A: Do you use life jackets? Any rescue equipment stock?

K: No, don't say stock, we seldom see and never use life jackets.

A: So, what do you wear to evacuate?

K: No, we just go up to the upper level, see the water and wait. We have rice, oil, and noodles stock.

These examples show that they understood their situation very well. They assessed the condition and knew their options. Their evacuation plans were based on their actions in earlier floods. They agreed that the best evacuation place was the roof of their two-floor house. They brought food that was easy to eat, such as instant noodles and bottled water. The women seemed to have a better command of food details than men did, but the men knew more about the supply of cigarettes. If the flood had not reached the roof but lasted more than 2 or 3 days, then the head of the RT would ask young men to look for logistical support using the plastic cart or a big plastic bag.

However, this scenario plan is not the same as the evacuation plan that was promoted by local and nongovernment agencies (see Fig. 5.2). The standard operating

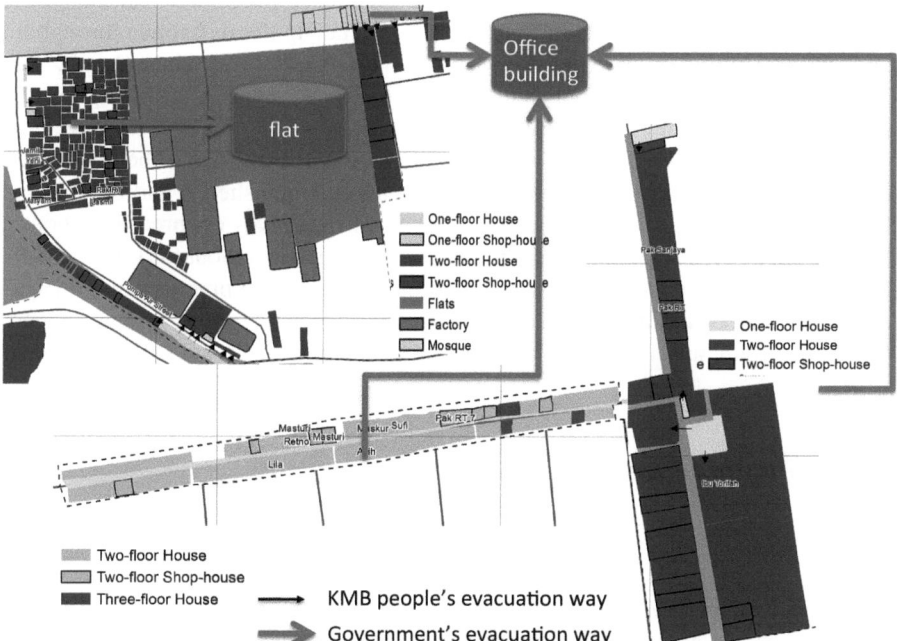

Fig. 5.2 Shelter and evacuation maps (Source: the author based on FGD RTs 7 and 15)

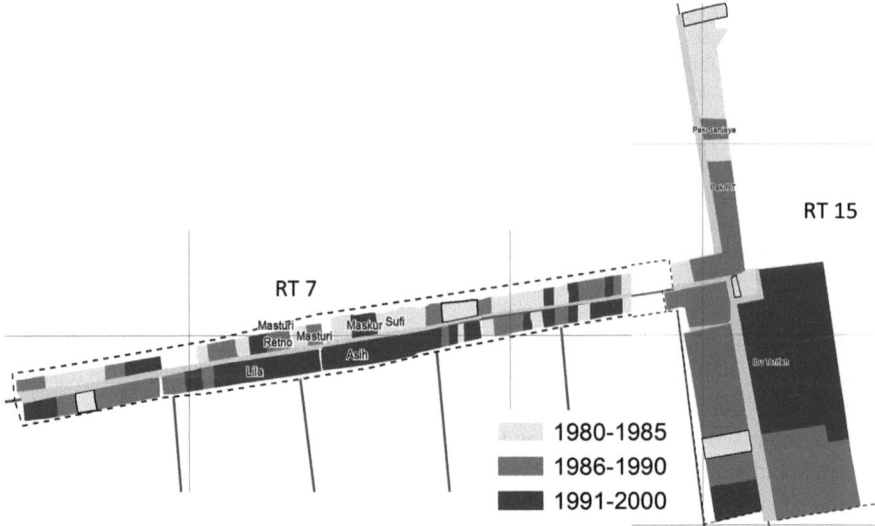

Fig. 5.3 Distribution of a two-floor house (Source: the author based on FGD RT 7 and RT 15)

procedure (SOP) in a flood emergency response requires them to evacuate immediately to the government shelters. However, this procedure applies only if the government has decided that the flood is disaster. It is thus problematic because for the KMB residents, a flood is a commonplace event. The big floods happen throughout the Jakarta area every 5 years. The annual tidal floods bring chest-deep water. Therefore, because the SOP only applies in major floods, it is not surprising that the shelters are located in the safest place in the KMB area.

Only the KMB people in RTs 7 and 15 used the government's shelter. On the north side of KMB, the shelters are four-floor flats and the fisheries harbor building. In most cases, people just climb onto the roof of their two- or three-floor house. Most residents of RT 7 and RT 15 have two- or three-story houses (see Fig. 5.3). Therefore, their evacuation plan was based on self-reflective practices and on self-organized and communal management, whereas the government's plan concentrated on certain buildings and was managed by the organizational structure.

5.1.2 Planning for Living Space Arrangements

A living space arrangement is a type of adaptation planning that belongs to individuals or households, but the planning of a house arrangement is constructed according to their shared understanding of houses that have been flooded. Many interviewees are women who stay at home. As in other *kampung*s in Indonesia, women are expected to take care of the home, but some of them also work

5.1 Reflection on Experiences and Learning

Fig. 5.4 Adapted one-floor houses (Source: the author)

informally for extra money, such as in open *warung* (small booth), doing washing, and as part-time servants. The housewives know what will happen to their homes if they are not well prepared. The embeddedness of their planning knowledge is generated and consciously shared among them in casual meetings at *warung*, *pengajian* in houses or mosque, or *arisan* (regular social gathering) and at neighborhood (RT) meetings.

Based on the interviews and group discussions in RT 7, RT 15, and RT 20, KMB people managed the room function by following what the neighbors have done before. They planned the houseroom management by reflecting on previous experiences of flooding and then learning from each other. One-floor houses had no upper floor or balcony. The owners did not have any plans to change the layout, but they installed a storage facility and raised the *teras* (the front floor) and floor to minimize the amount of floodwater that got inside. They used wooden scraps to install the storage, and they used cement and tiles to raise the *teras* (Fig. 5.4).

The storage facility was created on a hanging floor, which was nailed to the wall and used to store electrical and cooking equipment. The storage place was generally high on the wall. The *teras* was made of concrete, and it was at knee height. The floor was gradually raised above the level of the previous inundation. However, they could not raise the ceiling or roof. The houses are therefore becoming smaller instead of larger (see Fig. 5.5).

This knowledge of spatial arrangement was generated in the informal conversations among women. They had the same interest in managing their households.

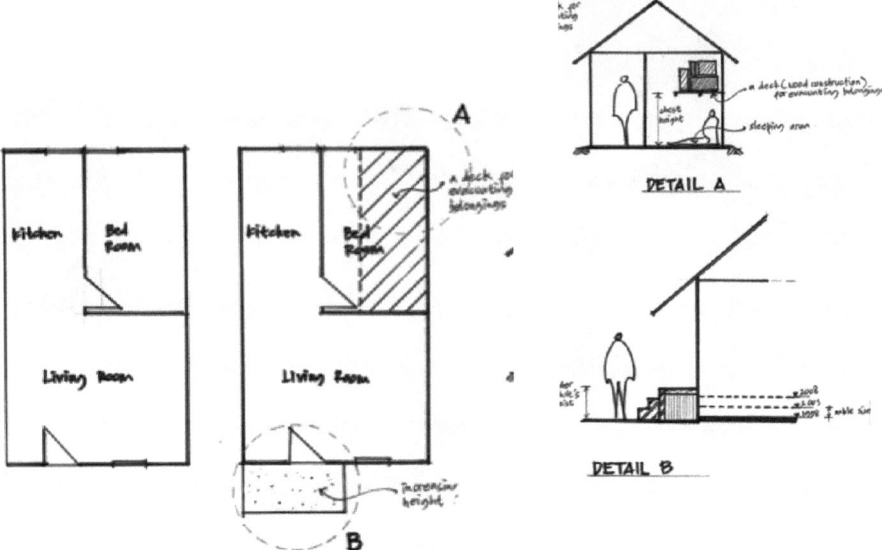

Fig. 5.5 Adaptation plan for one-floor house (Source: the author based on FGD RTs 7, 15, 20)

Although they seldom talked about how they managed their houses, they discussed the impact of the flood on their houses. They grumbled about the mud on their furniture, pots and pans, and appliances. Afterward, they paid more attention to the stories of their friends' who had solved these problems. Finally, they shared some ideas about storage.

In the group discussion in RT 7, Ibu Asih claimed that some of her neighbors had emulated her hanging cabinet. She said that her husband built a hanging cabinet in their living room to store their papers, flatiron, rice cookers, and flatware. Some of her neighbors asked them about the hanging cabinet when a *pengajian* event held in her house. Her friends asked how it had been made and why the height of the cabinet was only at the neck of an adult. She explained that it was because in her experience, the height of flood would never reach that position except the big ones when it would be easier to move her belongings on to and from the cabinet and move back and forth within her narrow place. She decided with her husband—learning from annual tidal floods—that they needed a hanging cabinet to prevent the hard work of cleaning their belongings after the flood had subsided. When I asked several women how often they emulated their neighbors, they replied that it was the easiest way to solve the same problem. The fragment of my interview to Ibu Asih on 25 June 2012 is a supporting argument:

> (I interviewed her after the group discussion).
> I: … Do you feel safe with the floor raised?
> A: Not really because the height of the *teras* floor kept the water back for just a short time. The water still entered my house. Our equipment got wet and was ruined. Therefore, we need the hanging storage place [she pointed to the wooden box hanging on the wall], at least until the water dried up.

5.1 Reflection on Experiences and Learning

I: I see… but how did you get the idea?
A: Ah… [She smiled] *ibu-ibu* [women] here already knew it. We often talk with each other [she used the bahasa term "*curhat*," which means a private talk].
I: So, [do] you mean that you got the idea from your neighbors?
A: Here, we are like a family. So we always share if we hear or have something new or even bad stories.

The women of KMB are very close. The term *curhat* (tell the story deeply) is used between trusted friends. Ibu Asih considers her neighbors "family." Her response also confirmed my observations in group discussions held in RTs 7 and 20. I could see from the clothes that that the women wore and from their conversation that they shared their solutions.

Two-floor houses are often used as temporary shelters for neighboring families. Most of the residents of both RTs built modest two-floor houses, each of which was about 18 m². The women said the ground floor was the main floor and the upper floor consisted of children's rooms and rooms for rent. During a flood, they would empty the first floor and move everything upstairs. They used the second floor for storage, sleeping, and neighbors that were guests. They just unfolded plastic carpets to gather their belongings. This functionality also applied to three-floor houses. A wood ladder gave access to the second floor. Some of them put the ladder along the outside wall if there was not enough space inside (see Fig. 5.6). Some people put the ladder inside the house, attached it to the wall, and cut a hole in the ceiling to access the second floor (see Fig. 5.7).

Fig. 5.6 Adapted two-floor houses; photo taken by the author

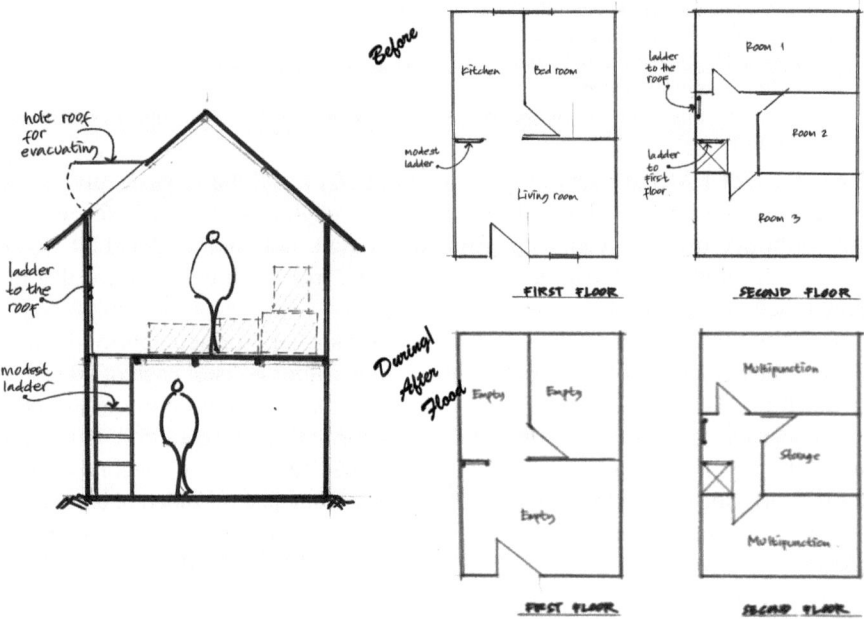

Fig. 5.7 Adaptation plan for a two-floor house (Source: the author, based on FGD of RT 7, RT 15, and RT 20)

The options for adding a ladder and using second floor as a multifunctional room were reflected in the experiences of what neighbors had done. In the interview with Ibu Maskur (RT 7), she said that her husband built the second floor as a bedroom for their four children. Before two of her children moved out, they started using their children's room as a storage area (interview with the author, 18 June 2012). Pak Parlan (RT 15) said that he added a second floor so that he could rent it, but since the renters always went back to their home village, he used the second floor to store his belongings (interview with the author, 31 January 2013). Even though second and third floors were added for reasons that have nothing to do with floods, they could be used for storage during a flood. For them, raising the floor is a way to accommodate a growing family and to generate rental income. The flood does not force them to change their house. They use the extra floor to store equipment, install a ladder, and put a padlock on the door. Therefore, I deduced that they planned the functionality of the rooms in their houses to adapt to the floods.

The interviewees commented that this living space management resulted from informal conversation, and it was not intentional. The men discussed it casually, especially at *maghrib* e. On this occasion, they talked about other neighborhood issues. Entong (RT 7) said that he often got useful information when they gathered in the mosque before their *sholat maghrib* (*maghrib* praying), which included the measures his neighbors took during the floods (interview, 23 June 2012). This small forum is an effective way to socialize and learn about what others are doing. This sharing process is analyzed in the next chapter. However, it can be deduced from the house management plan that it is based on common concerns about minimizing losses and damage caused by the floods.

With regard to house management, KMB people do what planning professionals do. The purpose of layout planning is "to determine the best physical arrangement of resources within a facility" (Reid and Sanders 2010, 3). KMB people planned to add storage that could be used in a flood. Therefore, the arrangement focused on increasing the functionality of their houses. KMB people followed the six steps of layout planning: (1) defining the objectives; (2) identifying main and supporting activities; (3) determining access, arrangement, and flow; (4) determining space requirements; (5) designing an alternate layout; and (6) evaluating and choosing a layout.

The objective was to protect valuable equipment. By identifying the main and supporting activities, KMB people decided that the primary function of the second floor would be economic and the secondary function would be storage (safety). In determining access, arrangement, and flow among household activities, KMB residents added a ladder and installed the storage that was accessible. In determining the space requirement, they also designated the room for specific uses. In designing alternative layouts, they used several imaginary sketches that were orally described. Finally, in evaluating and choosing a layout, they reflected on their knowledge and their neighbors' actions. Therefore, compared to the zoning regulation and building codes of Jakarta, their plans did not meet the standard.

5.1.3 *Planning for Neighborhood Infrastructure Provision*

The third case concerns the supporting facilities used to adapt to the floods in their neighborhood. KMB people adapted the house's surroundings in order to diminish the flood's impact on their belongings. They did not pay much attention to the house itself. Therefore, they created an adaptation plan that kept as much mud and garbage as possible out of the house.

They focused on improving the drainage and footpaths. They repurposed the drains not only to improve the flow of water but also to filter garbage. They attached a fisherman's nylon net and placed iron pins taken from construction equipment at the lower and upper drainage ends that crossed their neighbors. They used beach sand to build a footpath to and from their neighborhood. They use used wood and bamboo construction along the pathway to ensure that the sand sacks were strong enough to walk on (see Fig. 5.8).

Before the sea dike was built in 2007, KMB residents sandbagged the shoreline. Afterward, they build a barricade of sandbags to keep garbage from floating into their houses. Based on their experiences, men took the sand from the beach, and women poured it into the rice bags, which were supplied by the KMB secretariat. The height of the stacked sandbags was determined by discussions held by the heads of RTs with the neighborhood. The men, most of whom were construction workers, did the building.

Fig. 5.8 Adapted settlement infrastructure, photo by the author and RW 17 collection

The adaptation of the infrastructure reflected the emergent need to support their ingress to and egress from the neighborhood. They improved it from time to time based on the lessons learned from the previous flood. They counted on the availability of local resources. There was no fixed decision-maker; they made the easiest and least expensive decision.

On 24 June 2012, I talked to Ibu Entong in RT 7 about the decision to make the road.

> A: But, how does the road in front of your house? Do you help raise the road?
> E: It comes from our own sources, *swadaya* (a self-help mechanism).
> A: How many times have you raised it?
> E: *Wah*... many times... countless.
> A: How high do you think?
> B: Since I was here (1985)... [She thought for a moment]... I think more than 1 m ... because my house floor is still under the road level. While I have raised it about 70 centimeters.

I also raised this topic with Pak Kadir, the head of RT 7 on 7 July 7 2012:

> A: Was the flood level reach this level? [We stood on the alley road of this neighborhood, in front of his house].
> K: No, it has risen a lot... I think about 2 m. Before, there was a fence that the height is 2 m, now the fence already covered by the road. [He means the height of the road and fence is equal].
> A: Where do you get the funding? From kelurahan [government]?
> K: Hmm... just once. Mostly come from our people.
> A: How do you raise it up? What kind of materials did you use?

5.1 Reflection on Experiences and Learning

K: Generally, I invited two or three people here and discussed about the damage of our road. We also evaluated the last flood event, especially the height of the flood. Then we tried to calculate roughly how high we should raise up. Normally, we often raise 30–50 centimeters only, because we also count on how much debris from former buildings we can store.

A: Did you buy it?

K: Yes, but sometimes from the building demolition around here. We borrow a pickup if the location is far or a wagon if the location is nearby.

A: The manpower?

K: We did it ourselves. Each household raises the road in front of its own house.

A: Are people willing to do so?

K: Yes, they do. They use this road every day, don't they? So, they certainly will work in raising the road.

The adaptation planning agenda is derived from the head of the RT and anyone else who takes initiative. If someone needs something done, he proposes it to the head of the RT or tells friends and neighbors. For example, when they do *kerja bakti* (working together to clean up the neighborhood), they sometimes discuss it first. Because they attend *kerja bakti* every Sunday at the beginning of the month, the issues have always been updated. *Kerja bakti* is a ritual for planning adaptation to flood. It is not only a duty as argued by Perkasa and Hendytio (2003). The memory of drains clogged with garbage and the large number of mosquito larvae are the most general drivers of KMB people to conduct *kerja bakti*.

The neighborhood infrastructure planning also preceded through the reflective practices of KMB people. Kadir, the head of RT 7, wanted the road surface in his neighborhood to be raised so that it would be easier to use. Based on his knowledge, he assessed the possible height of future floods. He also searched for information and calculated the cost of the repair. He gathered the men in this neighborhood association to work together. KMB people did the same thing when they installed nets as disposal filters for the drains near their front doors. They remembered the stinking garbage left after an earlier flood. These planned adaptations resulted from their acquired foresight, which was developed through reflection and practice.

Therefore, the adaptation planning by KMB people is based on their memories of and information about the worst floods. They borrow tools and exchange information about modifying their houses. They generously share information about the solutions that they have found, but sometimes they seem to brag. Furthermore, although there are residents of different ethnicities in each neighborhood, they are all urban migrants. They see their neighbors as family, which makes it easier to share information about how to protect their houses from floods. The modality of lived experiences is a key factor in LEAP.

As these three cases showed, the application of reflective practices in adaptation planning responds to several issues pertaining to the adaptation. First, the adaptation of the poor is not only spontaneous but also planned. The KMB case study showed that the two are linked by reflective practices. Adaptation action, which is perceived as spontaneous, is modified to reflect previous adaptation. The reflection depends on the structure of *lifeworld* of the residents. Therefore, the planned adaptation of the *kampung* people who directly experienced floods differed from the responses of the government. The NGOs simply analyzed the flood.

In contrast to the typology of adaptation developed by Smit et al. (2000) and Smit and Pilifosova (2007), which identified the different forms and types of adaptation based on the goals and the period of adaptation, I argue that the typology of adaptation is also differentiated by the lifeworld of the adapters. Someone who has lived through the flood makes a significant contribution to defining the form of adaptation. Different people adapt to a phenomenon in different ways. Thus, the lifeworld of the person determines the form of adaptation. The distinction between planned and spontaneous or passive and anticipatory adaptation has no relevance in terms of the self-reflective practices used in the planning process. The spontaneous action provides knowledge about the weaknesses and strengths of past and present adaptations. Therefore, they reflected on them in planning adaptation. The form of autonomous adaptation is used as a baseline against which the need for planned adaptation is evaluated. Therefore, spontaneous and planned adaptation is linked through reflective practices.

5.2 Reflective Practice as a Source of Planning Knowledge for Adaptation

Adaptation occurs in different "systems," "units of analysis," or "exposure units" (Carter 1996; UNEP 1998). According to Garzon et al. (2012), adaptation planning at the community level is still in the initial phase. However, numerous books and manuals have encouraged community adaptation planning (CAP). Most of this literature concentrates on participatory and procedural mechanisms (CARE 2010; USAID 2009). Previous studies have not considered the lived experiences of flood-affected people and their accumulated knowledge resources for adaptation planning. This practical knowledge could be a useful source for the CAP.

Some scholars have suggested that local knowledge should be integrated into adaptation planning (Adger et al. 2009; Turner and Clifton 2009). Sagala and Damayanti (2010) recommended that the incorporation of individual actions should be done through a community planning approach. In the transmission of adaptation planning, knowledge could change local behavior (Carmin 2009). External factors, such as professional networks and associations, NGOs, and consultants, often transmit ideas and knowledge to the local community. However, knowledge is not easily transferred to the individuals because the content of technocratic language has not been translated into their daily language and has not been communicated in their social world. Therefore, the locally embedded knowledge should be included in translation and communication.

The *kampung* people like KMB people who are still rooted in rural culture have communal values. The Round Table Forum of the Pacific Rim Council on Urban Development (PRCUD) Urban Planning and Climate Change in Indonesia, which was held in Palembang, South Sumatera in 2011, recommended that "Indonesia need to harness its long tradition of community participation and self-help to come up with innovative local solutions to integrate poverty reduction with climate change actions" (Rabe 2011, 37). The forum also recommended early engagement, the provision of relevant information and knowledge, and capacity building of vulnerable groups as the keys to the adaptation to climate change. Therefore, locally embedded knowledge is needed at the community level.

Locally embedded knowledge is different from the knowledge used to explain the production of local knowledge. Local knowledge is produced within a specific environment. It was derived from Lindblom's conception. It is relevant to particular experiences and is used to assert that the experts did not produce knowledge exclusively. Therefore, locally embedded knowledge is a form of knowledge that is "embedded in local history, local memory, and local network" (Gaillard 2010, cited in Piccollela 2013). In other words, locally embedded knowledge is experiential knowledge that is repeatedly deposited in the realm of individuals in a particular spatial, temporal, and social situation.

KMB people used locally embedded knowledge in three types of planning. In the context of planning, they did not engage in the kind of data collection and data analysis that professional planners conduct. They mapped the flooded area through a self-sensing process based on historical inundation events that had occurred in their *lifeworld* (see Sect. 4.4), instead of applying remote-sensing or other geo-reference tools. They then analyzed the adaptation options through inter-shared meaning production that was generated from a series of experiences, instead of expert knowledge. Finally, they conducted planning through a series of self-organizing events, instead of depending on statutory city planning.

The locally embedded knowledge of KMB people is similar to *metis*. This practical and experiential knowledge was conceptualized by James Scott (1998) in his book, *Seeing like a State,* to counter the dominance of scientific knowledge in improving human development. A locally situated environment that is a domain of human practices and experiences can form a *metis*. It resembles a language that is easily learned by rote. The dominance of the scientific community in producing planning knowledge is especially evident in adaptation planning, which is a relatively new field of planning practitioners. The locally embedded knowledge must not only provide new insights at the microlevel but also be used to derive a planning principle that is too general to be implemented at the microlevel. The *lifeworld* thus defines the kind of planning process that KMB people can imagine and practice.

The embeddedness of knowledge depends on the structure of the *lifeworld* and the knowledgeability of individuals. Knowledgeability is "an ability to know something about certain themes, [and] is different for the different members of a culture, and changes, as it is itself a social product" (Antweiler and Mersmann 1996 13). KMB people showed that their *lifeworld* had defined their ability to plan for evacuation, house management, and infrastructure provision. Their perception that a flood is not an enemy and their ability to adapt to and even benefit from flooding shaped their adaptation planning.

Their experiences produced the knowledge that the residents are surrounded by water and that there have always been floods. Instead of trying to prevent the floods, they preferred to live with them. This change in knowledge was used in adaptation planning. The structure of the *lifeworld* delineated their planning zone and narrowed the horizon of future conditions that they needed to achieve. This *lifeworld* also functioned as a reflected media that modified the present adaptation and allowed them to invent future adaptive actions. Therefore, the structure of the *lifeworld* determined the level of knowledgeability.

The changed meaning of floods framed the adaptation pathway of KMB people, who have adapted repeatedly to floods. These practitioners use the shifts in meaning to plan their evacuation, living space management, and nearby support facilities. Living with floods is their LEAP. It is a simple plan that comes from the reflection on their flood experiences and their present and/or future adaptation. The lessons learned from the series of adaptation actions have been accumulated, articulated, and preserved in their memory. They are then consciously used to decide the kinds of present and future adaptations to floods. The KMB residents also borrow from others and expand their knowledge, which strengthens their adaptation planning. This intuitive knowing is the main modality of increasing adaptive capacity.

The LEAP process elucidates the relationship between adaptive and coping capacity, which have been discussed in terms of disaster risk reduction and climate change adaptation. From the perspective of people-centered development, this relationship should be strengthened to increase the human capacity in responding to immediate shocks and potential stresses. The IPCC (2012) emphasized the distinction between coping and adaptive capacity:

> Overall, coping focuses on the moment, constraint, and survival; adapting (in terms of human responses) focuses on the future, where learning and reinvention are key features and short-term survival is less in question (although it remains inclusive of changes inspired by already-modified environmental conditions). (IPCC 2012, 51)

The dimension of time and the type of responses are differentiating factors. In terms of planning, the two concepts should not be separated because the planning process departs from the present condition, reflects past experiences, and addresses future opportunities and challenges. Borrowing the learning loop framework of Argyris and Schön (1978), which is shown in Fig. 5.9, the LEAP of the KMB people is a continuous learning process that can be divided into three loops: the immediate response of KMB people to their first flood, the reframing process of a lesson learned from past experiences, and continuous transformations in individuals, collectives, and socially reflective practices.

In the single-loop learning phase, adaptation is focused on ascertaining whether the reactions are adequate and effective to reduce the loss and damage caused by floods. Single-loop learning focuses on preliminary actions that are based on present knowledge. It is similar to experiential learning in the transformation process of social–ecological resilience (Folke et al. 2009). In double-loop learning, the evaluation is extended to assess whether actors are "doing the right things" (Flood and Romm 1996), that is, whether the evacuation and house management have solved

5.2 Reflective Practice as a Source of Planning Knowledge for Adaptation

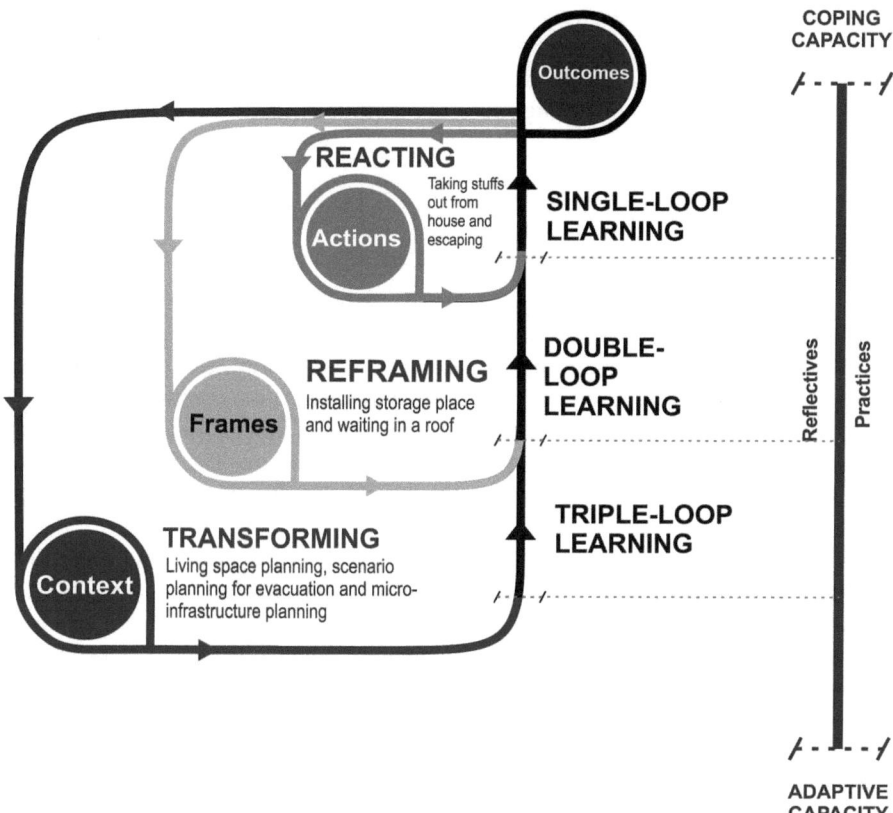

Fig. 5.9 Learning loop of KMB people (Source: the author reproduced from IPCC (2012), Pahl-Wost (2009), Stermann (2006), and Agrys and Schön (1978))

the real problem. Corrective actions are taken after the problem is reframed and different planning goals are identified (Pelling et al. 2008). In triple-loop learning, the actors question how the *kampung* and external institutions, such as the government, support their adaptation planning. It provides a domain and framework in which LEAP operates. Providing a planning framework for integrating double- and triple-loop learning, particularly by flood-affected KMB people, is more effective than narrower planned approaches, which depend on specific future climate information (McGray et al. 2007; Pettengell 2010). The three-cycle phase will strengthen their capacity, from only coping to have mode adaptive skills in responding the floods. Those who have already transformed their skills from cope to adapt can be named as the adapters.

LEAP is needed because the response to floods is perceived not only as a static reaction to a hydrometeorological catastrophe or a natural disaster but also as socially constructed by the affected community. Therefore, the *lifeworld* of the affected people should be discovered before linking the different actors in adaptation planning. Limited interaction results in planning divergence. It cannot consolidate all resources to adapt to the disaster events or climate change impacts. The KMB people used reflective practices to conduct their own planning process through informal forums instead of through formal institutions that were facilitated by the government.

5.3 Adapter as an Informal Planner

The informal planners in KMB are adapters who reflect on their experiences and those of their predecessors to make and enact plans before, during, and/or after flood events at different levels of the operational zone. At the community level, the heads of the neighborhood associations are informal planners, which is significant because they offer advice to families on adaptation planning. The KMB case study also showed that they preferred to use the existing forum to present or exchange planning ideas instead of holding special events.

The role of informal planners depended on where they operated, but it was limited by time and space. From the individual zone to the family zone, they tended to be resources for family or colleagues in providing the best practices for adapting houses and livelihoods. The informal planners liaised with other KMB stakeholders, such as company owners, harbor administrators, and the city government. They promoted planning idea at the *kampung* level. Therefore, I will now discuss the roles of informal planners for individuals, households, neighborhoods, and the *kampung*.

As stated in Chap. 5, adapters used self-reflective practices in making an adaptation plan to improve their livelihood during a 5-day flood. Even more at individual level, Fendi, who grew up in RT 16, earned money during the floods. He said that furniture and plastic materials floated around the neighborhood. In the first flood, he collected two rice sacks of plastic bottles. In the next flood, he persuaded his two friends to build a cart from debris. In addition to collecting plastic bottles, they used it to transport people (Fig. 5.10). Fendi's experiences showed that he could learn from experience and generate income without having formal knowledge. He framed the flood not as a problem but as a potential source of income. He honed his ideas to create income from the flood, improve the performance of his carts, and enact the same procedure with his friends. By using their practical knowledge, he and his friends profited from the floods. What Fendi and the others did is actually a planning process using knowing in action. Schön (1982) stated that the theory in use is embedded in the logic of actions. James Scott also refers to this unwritten planning knowledge as metîs in his book, *Seeing Like a State*.

At the household level, during a flood, family members often discussed the self-reflective adaptation practices that they or their neighbors have already applied. It was common knowledge that the worst flood is the benchmark for house management. The women were the resource people since they had the greatest number of stories. The meaning shared among the family members was used as a reference for

5.3 Adapter as an Informal Planner

Fig. 5.10 Examples of activities of individuals that have generated income during the flood (Source: the author, 18 January 2013)

the next flood. As in other *kampungs* in Indonesia, women took care of the household equipment. Since they adapted when flood comes, they knew better how to manage their houses. They learned to do better from their previous experiences, thus following the triple-loop learning process. Therefore, they showed intuitive planning that allowed them to adapt easily to the floods.

An example at the neighborhood level shows the evacuation planning process. For instance, in RT 7, there is a common understanding that when the water reaches chest height, the people of RT 7 will move into a two-story house. This is not the plan suggested by the local government and NGOs. People were required and trained to run to the shelters that had been decided as the logistic center, but most of them never did. Their evacuation plan reflected lessons learned from earlier floods. Unfortunately, they never mapped or wrote down the plan. They preferred to talk it over because they thought that reading and writing were not the habitual actions of KMB people.

At the *kampung* level, they built a footpath on the permanently inundated road on *Perindustrian* Street. This path enabled KMB people to enter and leave their neighborhood, especially from RT 15 to RT 20. However, because the government and industrial companies had never helped them to build the road, the residents did it themselves. They started to discuss it with Agus, the head of RT 15 in a *maghrib*. He insisted that they raise the inundated road to facilitate the movement of the residents. After two or three talks, they asked the head of KMB to solve this problem. The head of KMB tried to discuss it with the companies and industries on the harbor but failed. Then they decided to use beach sand and a sand fortress to make a footpath, and they asked the residents to help with the construction. As mentioned in the previous section, they organized themselves to implement this small road development through the spirit of *gotong royong*.

Therefore, the role of the adapters as informal planners is very important in the transmission of knowledge. KMB people looked to the head of RTs as the informal planners because that person usually had more information and experiences related to the floods. I observed that most interviewees mentioned the head of the RT when I asked who their resource person was during a flood. The heads of RTs also facilitate communication between households and the *kampung* secretariat. However, other resource people included Konedi at the *kampung* level, Bu Maskuni at the household level, and Fendi at the individual level. They preferred to share their knowledge personally with neighbors and friends. When the plan was workable and useful, the others applied it to their own houses.

In contrast to formal planners who make a sophisticated plan in order to convince the community, informal planners develop such plans in their everyday lives. This is an embedded process instead of a project-based activity. The informal planners use social events to share their plans and with their followers. Since it is a reciprocal process, the planning knowledge that comes from the informal planners is not exclusively separated with their followers' realm. It is a locally situated form and forming process. Therefore, the control resides in the community, whose members have the same rights to evaluate or modify the plan. The control mechanism is embodied in their informal social events, and it depends on the initiatives taken by the people.

At the community level, adaptation planning does not depend on the technical experts mentioned in the previous chapter and follow the emergent shift in the planning paradigm from participatory to people centered. The LEAP is one example of people having their own plan to adapt to their changing environment. Therefore, the process of adaptation planning at the community level should reflect the lived experiences of the affected people.

The case of KMB demonstrates that planning practices were generated by their habitualized actions in adapting to regular floods. It supports Schön's (1987) assertion that "the experiences (theory-in-use) that were embedded in the logic of the action" could be used in creating solution (cited in McDowell et al. 2007). Although the domain of their thoughts and actions were not sufficient to deal with floods, they constructed their adaptation planning even though the planning outputs were not available for long-term actions and strategic visions. Their *lifeworld*, which confines their framing in space and in time, explicated their capacity to develop

5.3 Adapter as an Informal Planner

adaptation pathway. This world must be recognized, before it can be connected to the larger world of the city.

The importance of the personal scale in the planning process has also been discussed in the recent discourse of planning practices (Chap. 2). Gehl's (2011) human-scale perspective and Friedmann's (1987) conception of territorial units of individuals have caused the urban planning discipline to examine the microscale and the subjective human experiences (Wagner 1970). Therefore, with the proposition that planning is a universal human activity, I argue that if the personal scale is taken into account, the urban poor, as represented by the *kampung* people, also conduct adaptation planning.

KMB people share their knowledge of adaptation planning by refusing to cast the flood as an enemy. They modify their houses and surroundings to reduce the consequences of flooding. They plan to live with the floods despite the inconvenience. The reflection on those experiences later shapes their zone of operation in adaptation planning. In addition, the lack of resources and/or isolation from formal institutions has driven them to concentrate on their immediate surroundings. Therefore, they engage in a "locally situated form of knowledge" (Antweiler 2004) that is similar to the metîs (Scott 1998).

The planning approach used by KMB people is based on their reflections on their previous experiences and understanding of the history of floods in their area. They do not use that knowledge to formulate a long-term solution. The LEAP of KMB is not preceded by a participatory process but is conducted by the construction of intersubjective meaning (Schütz 1967), which is intentionally shared in habitual actions (Berger and Luckmann 1967). The LEAP in KMB also showed that the adoption of Schütz's *lifeworld*, which elucidates the institutionalization of planning for a personal good, begins with the inner experiences of the people involved in the planning process. The *lifeworld* analysis revealed the institutionalization of adaptation planning in social construction because the *lifeworld* provides the context for and the domain of adaptation planning.

The KMB people used informal community events, such as *maghrib* talk, *kerja bakti*, and *arisan* to transmit planning ideas or implementations. In the process of LEAP, they used trial-and-error methods to actualize their locally situated knowledge. The lessons learned from previous floods were an important source. They did not realize that they had conducted a planning process, but they understood the steps used to design their evacuation plan, their house arrangements, and neighborhood infrastructure.

Living with floods required LEAP that should be understood as the real world of KMB people. They used locally situated knowledge (Antweiler 2004) that was formed through the exchange of habitualized adaptation practices to mitigate the effects of flooding and modify their houses and neighborhood infrastructure. In their LEAP, the KMB people have institutionalized the plan of living with the floods that occur in their neighborhood. This chapter discusses how they share these plans, establish "a rule of the game" in adapting to the flood, define their roles, develop control mechanisms, and reify the meaning as part of an institutional order. The social world is the arena where the function, practices, and repetition of the planning

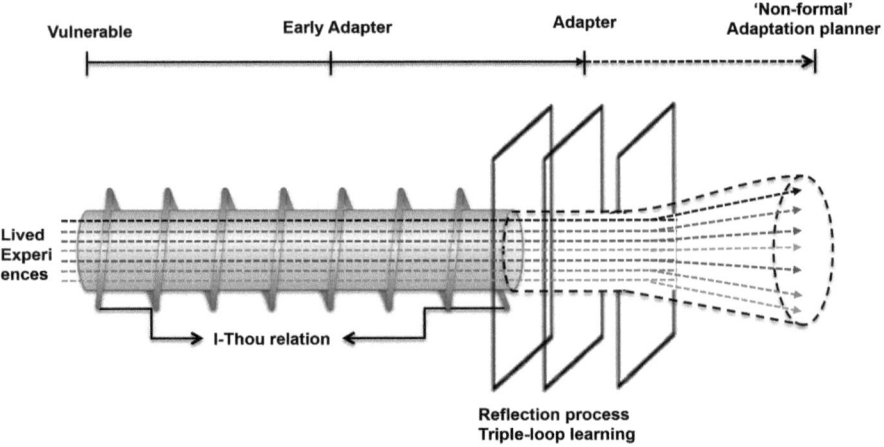

Fig. 5.11 Transformation process from vulnerable to adaptation planners

are decided. It is important to understand the KMB people as a typical of the *kampung* people in Jakarta. They are urban migrants who work in informal sectors and live in informal settlements, but they still practice their village traditions. Even though *kampung* have a symbiotic mutualism with the city, *kampung* people are different from urban people. The *kampung* people tend to think and act more covertly than other urban dwellers do. This informality also applies to KMB people in their everyday lives and in adapting to floods. Therefore, the reification of LEAP cannot be separated from the *kampung* as a community organization that is socially constructed by its residents.

However, the planners' perception on the role of urban poor is that they are only gatekeepers, participants, or contributors. They are seldom regarded as having their own planning knowledge. This research discovered that the KMB possesses knowledge of locally embedded planning. Extensive flood experiences provide a lesson learnt through self-reflective practices. The triple-loop learning cycle facilitates the transformation of KMB people from know nothing to plan something, such as experienced by Fendi, Agus, or Konedi (see Fig. 5.11). They could take a role as a planner when the situation needs them to do so. Identifying the type of people who has typical knowledge is very important in institutionalizing the adaptation plan, because, in fact, they have already initiated the planning process and adapt to the environmental changes, particularly in debating and selecting the best way to adapt their settlement to the flood.

In the context of building resilience, some economists believe that urban poor are vulnerable. However, not all vulnerable are poor in terms of social capital. Moser argued that "individuals in poor communities often have a great deal of social capital: informal reciprocal relationships between individuals and families...such as community organizations" (Moser 1998). Martin-Breen and Anderies (2011) added that:

5.3 Adapter as an Informal Planner

> The poor are not helpless; that is, they already organize to build resilience. Provision of only direct solutions to poverty—building new housing, providing employment, health care services—may indeed reduce adversity, and even move someone out of poverty, but they do not necessarily build resilience; it may even erode the existing resources that contribute to it.... (Martin-Breen and Anderies 2011: 28)

"The capabilities of the poor population to use their resources to reduce their vulnerability" (Moser 1998: 14) need to be addressed by urban institutions. However, only few cities have a policy framework that enables endogenous knowledge of resilience, such as owned by urban poor, to grow. Since these settlements are located in a flood-prone area, consist of temporary buildings in an irregular pattern, and lack a basic infrastructure, the municipality cannot always support the poor communities in adapting to the flood. Therefore, building urban resilience needs to consider the existing or endogenous sources of resilience.

The levels of planning discussed in the previous section show that the *lifeworld* of flood-affected people has shaped their planning way. Alfred Schütz states that the individual's *lifeworld* depends on and is determined by his or her position in time and space (Schütz 1967). From the individual to the *kampung* level, no provision in those planning processes is provided to move them from the flood-prone area or relocate them to safer places. When I apply the *lifeworld* analysis, I see that living with floods is an aggregate of what KMB people know about flood. It is not a reactive emotion of KMB people, but a reciprocal meaning that constitutes after the long reflection on the adaptation options that they have taken.

The *lifeworld* analysis also helps identify the reciprocity of women and men in the social events of KMB. The social world of KMB people is the domain for planning. The presence of adapters, who can survive and even benefit from the floods, is central in the production of planning knowledge on adapting to floods. For example, staying on the roof is knowledge that resulted from the shared meaning of reflected experiences. According to Agus, the head of RT 15, this knowledge is perceived differently depending in a person's familiarity with it. He knows what he does with staying on the roof because he has a scenario to survive and escape if the flood reaches the ceiling. He argues that the lifeboat and any logistics will be easier to reach from a roof. It is a designated plan. He realizes that this knowledge has been socially distributed but still depends on the people themselves. Some people may know, but if they see it was not relevant to them, they do not stay. The others probably do not know at all because they do not care (Field note, January 2013).

The other example is the knowledge of adding a ladder to two-floor houses. Kadir, the Head of RT 7, reminds people who add a second floor to put the ladder in front of the house where it is easier to reach, not inside. Several accidents occurred because of the small size of the house. It thus becomes more dangerous for people to bring their belongings back and forth during flood time. However, he also told me that not everyone follow his advice because they think it is not relevant for their cases, but most of their neighbors have applied this knowledge (interview with the author, 22 June 2012).

The knowledge of plastic bottle collection during floods is another example of people emulating each other. As told by Fendi, the idea of collecting plastic bottles came from the collection of someone during the flood in 2007. He saw a 1 × 2-m net that is installed in front of *Gang 1* (alley) to catch the garbage brought by flood. He just memorized the feature of the net without asking the installer how to make it. He thus followed and developed further the net by making a capturing net in a 30-cm-diameter hole. He made a used wood grip to hold the net, like a tennis racket. He sat on a dry place in front of the mosque but sometimes waded into the water to catch plastic bottles. Two of his friends copied him; they earned money from the severe flood (interview with author, 28 January 2013).

Those cases support Berger and Luckman's claim that knowledge in everyday life is socially distributed, possessed in different ways by different individuals and by different types of people. It depends on the sum of the individual *lifeworlds* (Oberkircher and Hornidge 2011). Therefore, the institutionalization of the LEAP is more effective because the plan uses the language that originates in and has primary references to everyday life in the KMB.

The effectiveness of the LEAP's institutionalization can be also proven by considering the ignorance of KMB people regarding the formal plans and policies made by government and NGOs. Clearly, these planning processes and outputs are not congruent with the realm of *kampung* dwellers. For example, the relocation planning of 7000 households in the *Pluit* lakeside, including KMB dwellers, to new flats that would be built on 8.3 ha in the KMB did not suit the preferences of the KMB people. The KMB dwellers' preferences were different from the government's plans (Jordan 2013). The other examples are the differing responses to the floods. As explained before, the KMB dwellers preferred to wait out the flood on their roofs and not go to government shelters. They also used their own modest means to evacuate when necessary. Both responses showed that the domain of government planning did not recognize the adaptation pathway that had already been planned by KMB dwellers and was constituted in their community.

5.4 Institutionalization and Reification of LEAP

By using the chronological stepwise approach of institutionalization processes, which Peter Berger and Thomas Luckmann assessed in *The Social Construction of Reality: A Treatise in The Sociology of Knowledge* (1967), I describe the objective meaning of living with floods as the LEAP. This objectivation occurs because reality is an intersubjective interpretation:

> Everyday life presents itself as a reality interpreted by men and subjectively meaningful to them as a coherent world… [It] is a world that originates in their thoughts and actions, and is maintained as real by these. Before turning to our main task, we must, therefore, attempt to clarify the foundation of knowledge in everyday life, to wit, the objectivation of subjective processes (and meanings) by which the intersubjective common sense world is constructed. (Berger and Luckmann 1967, 19–20)

The previous chapter showed that reflective practice based on lived experiences shifted the focus of adaptation planning from evacuating to staying on the roof, from protecting the house to converting the space during floods, and from blocking the drains to filtering the garbage. Such changes resulted from intersubjectively shared reflective practices. These changes were developed in the course of informal social gatherings. These events were the main forums for the reciprocal typification of LEAP by KMB people. In the following section, I explain the construction of three cases of LEAP: typification, institutionalization, legitimation, and reification.

5.4.1 House Management Plan

Outside the flood season, house management plans are shared among the women in each neighborhood association in KMB. They usually talk about their furniture or the condition of their rooms. During floods, they usually talk about how messy their houses are and what they did. The houses in the mouth of the alley tend to stay flooded longer than houses in the middle or at the end. After the flood, they talk mainly about how they repaired the damage and what should they modify their preparations for the next flood. Based on a FGD in RT 7, I observed that they discussed what Ibu Ipeh, one of the participants, did to store her kitchen equipment.

The KMB people, especially the women, used LEAP as the "company of their operating procedures" in adapting to floods. The women carried out a house management plan. It was practiced repeatedly, and it reflected how they and their tenants adapted to the KMB environment by modifying their rooms:

> Living permanently in KMB is not cheap because we have to build at least a two-floor house like most residents here. Emptying a ground floor for floodwater is a model of all KMB house… [In the future], if I have enough money, I will build my house like that. (Nasir, interview with the Author, RT 20, 12 October 2012)

The habitualization of this house management plan compelled the KMB people to be innovative and critical. For instance, they added a ladder to reach the second floor (see previous section). They discussed whether to use iron or wood, whether the position of the ladder should be inside or outside the house, the number of stairs to install, and whether they should use odd or even numbers. The plan could be modified because the residents came from different cultures and understood floods differently. By having access to many examples in the KMB's houses, the KMB people optimized the functionality of their rooms. Hence, institutionalization had already occurred. Berger and Luckman asserted, "institutionalization occurs whenever there is a reciprocal typification of habitualized actions by types of actors" (Berger and Luckmann 1967, 54).

Having storage facilities on the upper floor or hanging on the wall of the ground floor allowed them to feel safer and worry less about regular floods. Because they benefited from this plan, it is not surprising that they adopted it and recommended it to their friends. I observed these plans being discussed in *arisan*, *warung* talks, or in sidewalk chats. Even though the topic was not always about this plan, many participants raised this issue, especially when they saw it on television during the rainy season.

I observed that, although the *arisan* was often combined with *pengajian* (*Al-Quran reading*), the housewives discussed their everyday lives before and after *pengajian*. They also talked with each other after returning from the market in the morning or in the afternoon after bathing time. I often observed them talking about the inundation and the acceptance of logistical support. I also often observed groups of men talking in *warung* and on the sidewalks in the morning before they went to work and at night after dinner. They talked about the news and about problems in the neighborhood. There were several conversations about the flood, such as the inundated road in front of the fishery port, the condition of their houses and the mosque, and evacuation roads.

Because KMB people have informal social events, the information pipeline is informal. The secretariat of the KMB had administrative functions and facilitated social relations among heads of RTs, religious figures, and CSO leaders. They often used the secretariat to discuss strategic issues. This circumstance allowed the continuous transmission of knowledge. KMB people used both forums to socialize and internalize problems and solutions. Using part of the house for storage had become a rule of thumb for KMB people. However, there was no penalty for not doing so. They knew that the people were responsible for their own decisions:

> For your own good, I think [we] must have upper floor for keeping our equipment out of flood. Here, building two-floor [house] seems like a must. Many of us did just like that. If not, your furniture equipment would be easier to be broken. (Cicih, interview with the author, RT 15, 18 October 2012)

Therefore, everyone who built a house in KMB must have a storage room. The reification of this plan occurred through intensive social association. Whenever a house was built, either the head of the RT or the closest neighbors recommended safe storage facilities in case of flooding. For Agus, the head of RT 15 and the son of the former head of RT 15, this precaution was embedded in his knowledge and that of his neighbors. As a second generation of residents in the KMB area, he reified the house management because he had already experienced the advantage of having storage facilities:

> For new people, if they want to build a house, they must think about using their second floor during a flood … So, they would not be rushed when time comes… For people who live here, they knew already…it's automatic…it is inside their head [smiling and pointing to his head]… therefore, new people should ask the old residents here. (Agus's statement in the FGD in RT 15)

5.4.2 Evacuation Plan

In flood evacuation, KMB people used the planned scenario of facing a flood situation (Table 5.1). The heads of the RTs institutionalized the scenario during floods because they taught their neighbors what to do. KMB people were more inclined to listen to the RT than to the rescue team or another organization. Staying on the roof

5.4 Institutionalization and Reification of LEAP

is the best option based on their experience. The success stories about people who stayed on the roof have influenced the formulation of the scenario plan. Consequently, they considered it guidance in taking an action during flooding time. They did it automatically. The heads of RTs have no difficulties convincing people to follow the scenario because the plan has been internalized.

The advice to stay on the roof was not only widely transmitted within the neighborhood association but throughout the *kampung*. According to Konedy, "This is [refer to staying in the roof] how we adapt to the flood situations" (Konedy, interview with the author, 12 December 2012). He emphasized that KMB people preferred to stay until the flood reached their second floor. The enactment of this knowledge was confirmed by Andre and his friends, who are volunteers with the *Palang Merah Indonesia* (Indonesian Red Cross). I interviewed them on 18 January 18 2013 (the second day of the January flood). He asserted that the behavior of KMB people in staying on the roof was well known by most donor agencies.

> We already know their choices… but we keep persuading them to evacuate in order to avoid the other potential severity. We always do our best, but they insist on staying… We just run the emergency response procedure for their own good. (Interview with the author, January 18, 2013)

The media reported that people were staying on their roofs during the January 2013 flood. At that time, online media reported that even though some KMB people chose to evacuate, many remained in the settlement. For example (see Fig. 5.12), on

Fig. 5.12 Staying on the roof (Source: Media Indonesia, 21 January 2013)

January 2013, *Media Indonesia* reported that "thousands of people of *Waduk Pluit*[1] refused to be evacuated." *Pos Kota,* the local newspaper, also reported that government staff and army had always persuaded local people to evacuate (Ilham 2013). Therefore, staying on the roof has been reified by the KMB people.

When I was there on the second day, some neighborhoods on the north side, such as RTs 15, 20, 21, 7, and 13, were flooded by water that was more than 1.5-m deep. Therefore, the residents took shelter in bus stations, police stations, the main mosque, and an elementary school. According to the data gathered in my interviews, men brought their wives and children to the shelters, which were less flooded than their homes. The men told me that they went back to sit on their dry roofs until the waters receded.

However, some groups took the bus out of KMB; the bus was parked in the main toll road, just 500 m from the mouth of KMB road. The private buses were lined up, while their drivers shouted for passengers, most of whom were renters in KMB. When I asked several of them where they were going, they said that they were headed back to their hometowns, such as in Banten area (2 h away), central Java (4 h away), or west Java (2 h away). They were leaving because they had no valuable possessions, and they were seasonal migrants in Jakarta.

Therefore, based on my observations and interviews during the 5-day flood, the permanent residents of the KMB were more compliant with the plan than the renters were. There were no sanctions for those who did not follow the plan, but they did not receive assistance when it arrived.

> If you live here, it's better to do whatever the heads of the RTs and your neighbors who have been here the longest. They know this area well…they know how the things are run…. If head of RT does not know you, we are not responsible if anything bad happens. (Interview with the author, Gustara, 21 June 2012)

5.4.3 Another Plan

Another example is the 2007 flood when KMB people built the sea dike. Sandbags have been used to build a sea dike since the flood in the early 2000s. Damin, a resident of RT 15, said that he used sandbags to protect the front of his house. He said that in every flood, his neighbors pile sandbags and that his uncle had asked him to buy rice sacks in the morning market. Therefore, during the 2007 flood, he remembered to use them. At that time, because of the scale of the flood, most KMB people helped build the sand dikes. Hence, the use of sandbags was familiar. Even the provision of rice sacks was coordinated by the head of the RW, and they were distributed by the heads of the RTs. According to him, this is common in the rainy season; many sacks are available in the market, and street vendors sell them (Field note, June 2012).

---

[1] Waduk Pluit is a big reservoir in the west KMB. The word sometimes is used to describe the surrounding community, including KMB people who live in RT 19, 12, and 21.

5.4 Institutionalization and Reification of LEAP

Table 5.2 The multilevel institutionalization process

Factors/level	Individual	Family	Neighborhood	*Kampung*
Method(s)	Self-reflective practice	Sharing time	Inter-shared meaning	Informally organized meeting
Substance(s)	Alternative livelihood	Houseroom management	Evacuation and shelters	Flood emergence response
Local actor(s)	Adapted people	Adaptive housewife	Head of RT	Secretary of KMB
Main role	Planning maker	Planning maker	Advisor and social mobilizer	Facilitator to the *kelurahan* office
Planning media	Private event	Daily interaction	*Warung* chitchat, alley talk, etc.	*Pengajian*, RW meeting, etc.

Source: the author

These planning initiatives also show that the KMB dwellers had informally and spontaneously organized the planning process (see Table 5.2). At the individual level, the adapters used informal sources, such as learning from colleagues, in their individual planning. At the family and household level, the housewife had several optional sources of knowledge, which included family or conversations with her neighbors. At the neighborhood level, the heads of the RTs usually initiated the planning process but informally through *warung* chitchat, *pengajian* in houses, *arisan*, or RT meetings. At the *kampung* level, the secretary of the KMB discussed the solutions that involved the neighboring RTs through an informal *kampung* meeting.

There were no specific meeting agenda, no written invitation, no precise time to open or close the meeting, and no minutes of the meeting. When I asked why they did not formalize the meeting, they smiled, and one participant simply said: "We like this way…." The secretary of KMB claimed that KMB people are accustomed to listen and talk, not read and write. I observed the same behavior in group discussions held in RT 7 and RT 15. The KMB dwellers have established an informal way of operating the planning process. Therefore, the informal spaces of planning are constituted by the *lifeworld* of the KMB dwellers. Ananya Roy stated, "engagement with informality is in many ways quite difficult for planners" (Roy 2005).

Although many cities have developed a flood infrastructure (e.g., sea dike, canals, and water gate and a flood warning system) to protect the living place, numerous scholars emphasize the importance of the human dimension in urban resilience, particularly on the poor. Participatory planning is clearly perceived as a tool capable of recognizing the presence of and absorbing the input from the community. Less clear is the role of the urban poor in using their own resources to make their settlement resilient. It is often unobserved since their planning knowledge is tacit and they are unfamiliar with the formal planning process. Therefore, the ownership of the planning process needs to be addressed (AKP 2013).

Ownership has a significant role in institutionalizing the plan. In the urban context, the literature shows that participatory planning is imperative in building systemic resilience to climate change (UNFCCC 2009) because it will make sure the local aspiration considered and meet the requirements of vulnerable groups (UNU-EHS 2011). It is driven by community with respond to location-specific needs (UNDP 2004) and use the shared learning dialogue (Tyler et al. 2010). The literature implies that the participatory planning model is a suitable framework for the incorporation of local knowledge.

Some international organizations have developed a community toolkit to facilitate the adaptation planning process, such as a CBA (community-based adaptation) toolkit (CARE 2010), a CRisTAL (community-based risk screening for adaptation and livelihood) user manual (IUCN et al. 2012), CoBRA (community-based resilience assessment) guideline (UNDP 2013), a guidebook of adaptation to the coastal climate change for development planners (USAID 2009), and an Adaptation Planning Handbook by Canadian Institute of Planners (CIP 2011). Planning knowledge is still driven by technically procedural planning, is oriented to the degree of vulnerability of the place, and depends on the competence of professional planners.

Planning debates adopting a people- or place-based approach are widespread. In adaptation to climate-related disaster, most studies have taken a place-based approach, derived from the notion of vulnerable place. The discourse on people-based approach still lacks empirical evidence; "the planning is a self-conscious universal human activity that involves the consideration of outcomes before choosing amongst alternatives" (AICP 2013). The poor are the most vulnerable when disaster comes since they are located in the least safe areas and lack resources (Satterthwaite 2007). Community-based planning should have a role in development (Stiftel 2000). Jan Gehl (2011) addressed the importance of shifting the planning paradigm to the personal scale. He suggested that transportation planners should look at the problem of the citizens' need not the cars.

Therefore, the integration of the human dimension into building resilience requires a different perspective on planning. The resilience planning that is driven by local people needs to be examined. The Commission on Climate Change and Development (CCCD) recommends encouraging local adaptation strategies, such as the point of departure for engagement, learning from past experiences, the development of adaptive capacity by the people, the promotion of locally owned capacity, the prescription on diverse solution, the promotion of ecosystem services, and the provision of public funding support for the poorest (Christoplos et al. 2009).

5.5 Connecting to City Level: A Language of Planning Knowledge Converter Required

The government of DKI Jakarta Province has attempted to integrate *kampung* into the city development as follows: (1) administering *kampung* as a part of community services unit, (2) establishing *Rembuk RW*[2] or *kampung* meetings as the first tier of

---

[2] According to Bappeda DKI Jakarta (2013), Rembuk RW is the community consultation meeting at the RW level held to identify problems, needs, and the activities in order to solve problems based

5.5 Connecting to City Level: A Language of Planning Knowledge Converter Required

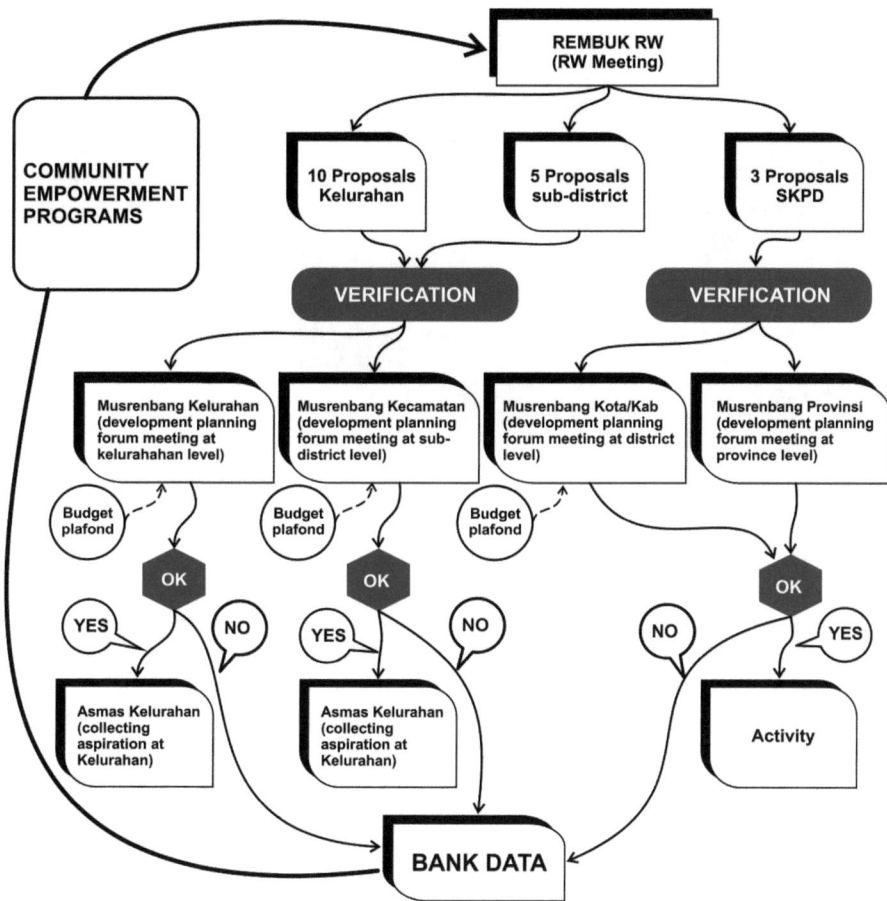

Fig. 5.13 The position of Rembuk RW in the planning system of Jakarta (Source: Bappeda DKI Jakarta (2013), translated by the author)

the development planning process, and (3) providing financial assistance for the heads of RWs and RTs in order to strengthen government services. Through the annual *Rembuk RW* (see Fig. 5.13), the government applies a bottom-up form of development.

Rembuk RW is a forum for *kampung* people to inquire about facilities, infrastructure, or any other support that they need from the government. The list of inquiries is submitted and discussed at *kelurahan* and *kecamatan* based on the scale of the program and the technical department based on the type of program. These offices select the program and propose it to the musyawarah perencanaan pembangunan

---

on priorities that should be formulated together. Previously, the head of the RW was a kelurahan representative at community consultation meetings. Since 2013, the consultation meeting has been conducted at the RW level.

Fig. 5.14 Rembuk RW 17 on 11 January 2013, photo taken by the author

(*musrenbang*) (development planning forum) at the provincial level. The selected program is implemented in the next fiscal year.

I observed the *Rembuk RW* 17 on 11 January 2013. I found that the consultation meeting resembled a one-sided conversation. The facilitators from kelurahan dominated the meeting by introducing a new glossary and practical information about *Rembuk RW*. The program ran from about 7:00 p.m. until 8:30 p.m. It was attended by 20 heads of RTs and three employees of the KMB secretariat. The head of KMB opened the meeting, followed by instructions from the *kelurahan*'s staff. Afterward, the meeting participants asked questions, and the meeting closed. However, at the beginning, the situation was formal, but after the opening speech, the meeting became more casual, and the attendees stopped paying attention (see Fig. 5.14). Only three questions were asked, and these did not seem relevant to the meeting. The people at the meeting did not seem to take the *Rembuk RW* seriously.

There are many hurdles to bridging these two worlds. One is that there was no meeting agenda. Two days before the meeting, in my conversation with Konedi, the secretary of KMB, I suggested that he propose the flood preparation to be on agenda of *Rembuk RW*, considering that it was January. He agreed and then proposed the flood issue to the floor. However, the *kelurahan*'s staff, who was the facilitator, did not agree to focus on the flood preparation. He insisted on showing how to fill out the form of *Rembuk RW* as shown in Fig. 5.15.

There was also a technical problem. As shown in Fig. 8.6, the *kampung* people were asked to provide the volume (column 6) of and the unit cost (column 7) of their proposed solutions. This format seemed to make it easier for the government to compile all programs from the neighborhood at the provincial level, but it would also limit the options to physical solutions. As a result, the KMB people complained that they did not know how to measure and calculate the volume or standardize the unit cost of their solution. They also struggled to understand the terminology used

5.5 Connecting to City Level: A Language of Planning Knowledge Converter Required

Problem Identification at RT level

RT. ....RW...............Kelurahan.....................Kecamatan ....

No	Problem	Causal factors	Proposed solution	Location	Volume	Standard	Priority numbers
(1)	(2)	(3)	(4)	(5)	(6)	(7)	(8)
1							
2							
3							
4							
4							
6							
...							

Stated at Jakarta, ...2014
Head of RT

(........................)

Fig. 5.15 Form of *Rembuk RW* (Source: Bappeda, 2013, translated by the author)

and to follow the framed format to input their aspirations. They did not understand the formal discussion of Rembuk RW.

However, they still filled the form of Rembuk RW, except the columns of volume and standards. When I asked Konedi, who collected the lists from the heads of RTs, he admitted that he also could not fill both columns because he also did not know how to calculate the volume or the unit costs of the solution. Consequently, he did not compile the forms of RTs into a single recapitulation form. He just gathered the papers into one envelope and submitted them to the *kelurahan* office.

According to Konedy, the secretary of KMB, Rembuk RW is a formality, not a solution. The questions he asked had not been considered in the replies from the *kelurahan* office. For example, he requested a garbage wagon. He has already mentioned that he needed a smaller wagon because of the narrow alleys in the settlement, but they sent big wagons. Therefore, the garbage wagons were left at the mouths of the alleys, which created the new problem that the garbage collectors were late. Another example is the request for tools to handle flood emergencies. He had made several requests for equipment such as lifeboats, big rope, lifejackets, and power supplies (*genset*). However, he has never received what he believes that KMB needs to prepare for the floods (interview with the author, 24 January 2013).

The head of KMB and most of the heads of RTs did not trust the *Rembuk RW* to meet their needs. The programs that they suggested in 1 year were not implemented in the next year. There were always changes or delays. There is no common planning language between KMB and the government. However, if a big flood occurs, the secretariat of KMB always coordinates with the *kelurahan* office, *Badan Penanggulangan Bencana Daerah* (BPBD) (disaster management agency of Jakarta), the police office, and the Indonesian Red Cross. Many companies, NGOs, and other agencies offer support but on an ad hoc basis.

The other difficulty in institutionalizing the government's plan is the relocation program. The relocation plan does not consider the realm of KMB people. In one case, it was even at odds with the KMB people's plan. For example, in 2007, because of the harbor expansion, PT. Pelindo, the state-owned harbor company, moved some residents of RT 15 through a compensation scheme. Only some houses were relocated, and the rest, 50 houses or 120 households, refused to move because they had evidence of their land status. The relocation of those who had the land status is still pending. In 2010, PT. Pelindo built the barricade of rock gate as a sea dike to protect the land from tidal floods and to separate PT. Pelindo's land from the 50 houses of RT 15.

However, the people of RT 15 covertly made a body-size hole at the rock gate in case they had to escape from a fire or other disaster (see Fig. 5.16). According to Komaruddin, one of the elders in RT 15 claimed that PT. Pelindo tolerated it for humanitarian reasons. Because there had been no further construction, they tacitly used the hole as a back door. Some residents used the space for water storage, raising chickens or livestock, and visiting neighbors.

Extending the back yard is a social problem. When I joined a group discussion, I saw many signs of PT. Pelindo's land across the backyard. According to Agus, Head of RT 15, those were outsiders who had started to build a temporary shelter and then

Fig. 5.16 The "back door" of the rock gate (Source: the author (2012))

5.5 Connecting to City Level: A Language of Planning Knowledge Converter Required 141

built a house. Most of the people plan to live there. Therefore, KMB people always reject relocation plans even though some programs were executed by the government. They do not understand the world that the government plans to build.

This unconnected world thus exacerbates the differences in planning between the government and KMB. Interaction between government and KMB has not occurred since the beginning of the planning process. KMB people believed that they could live with the shocks and stresses associated with floods. They insisted they could adapt. However, the government wants to relocate them. The result is a counterproductive plan.

The findings of the case study indicate that in order to institutionalize a planning product, the planners must understand the individuals' *lifeworld* from the beginning of the project. This first step is imperative in understanding the intersubjectively shared meaning in their social world. This meaning was already constituted, and they perceived it as reality. Planners then could develop a planning process that considered the language, interests, and relevance of the *kampung* people. This approach would address the problem of connecting planning to the reality of lived experience (Baum 1980).

In many cities in developing countries, including Jakarta, informality is embedded in urbanization. The notion of informality emerged in the 1970s. Many economists discussed the perspective of the informal sector in urban migration and employment. Nezar AlSayyad, professor of urban planning at the University of California–Berkeley, located the origin of urban informality in the literature of the 1970s in several studies, such as "State vs. Trade-service Sector," "Wage-based vs. Traditional/underemployed Services," and "Protected vs. Unprotected Labor Market" (AlSayyad and Roy 2004, 10). He argued that although urban informality does not simply relate to poverty or marginality, the informal sector, informal jobs, informal settlements, and informal markets are historically associated with poor communities. According to him, the linkage between informality and the poor was initially examined in relation to slums or squatter settlements, the shifting location of new informal settlements, and the gradual encroachment of public spaces. Thus, the role of informality may evolve depending on how it was constructed.

Informality has also been examined from the sociological perspective of institutions, such as North's (1990) distinction between "formal rules" and "informal constraints" (cited in Hornidge 2011). Many scholars have used this distinction to distinguish formal from informal institutions. Formal institutions are characterized by legal regulations, designed products, explicit rules, and law-court enforcement, whereas informal institutions are nonlegal or illegal, spontaneous, tacit, and peer sanctioned. However, according to Hodgson (2006), North's distinction failed to explain the difference. Moreover, the distinction between formal and informal institutions does not merit examination because there are still linkages or dependencies between them. Therefore, it is more worthwhile to understand how institutions are formed and sustained than to compare them.

As a social construct, institutionalization takes place through "a reciprocal typification of habitualized actions" (Berger and Luckmann 1967, 54). LEAP is a socially constructed product that has been performed by KMB people for more than a decade. The adapters of KMB showed that they could produce adaptation planning even without a formal knowledge of planning. They used the reflective practices based on their *lifeworld* as a flood-affected people who were also poor. Furthermore, they institutionalized LEAP into everyday life and proved they could live with floods.

In contrast, informal institutions are embedded in the culture of a community. As discussed in previous chapter, the *kampung* can be interpreted as an example of the informality in Indonesian urban development because "urban informality cannot be separated... [f]rom certain area-studies discourse" (Roy 2005, 155). The role of the *kampung* people should be central in disaster risk management because they are the first to be affected. They know better than anyone does about the local potential and limitations in reducing their flood-related vulnerability.

As discussed in Chap. 3, Jakarta's response to frequent flooding is aligned with the evolving institutions of flood management. The changes in institutions are evident in the flood infrastructure planning process, which has evolved from a structured plan to a management plan, from local to global cooperation, and from controlling to utilizing floods for development. However, Jakarta still prioritizes flood mitigation through the development of infrastructure instead of community empowerment. In contrast, United Nation Department of Economic and Social Affair (UN DESA) (2003) has recognized the value of community-based forecasting systems in reducing loss of life and property. They also suggested using local empirical knowledge and increasing local capacities to reduce communal vulnerability to floods.

As explained in the previous chapter, the KMB people based their adaptation plans on their locally embedded knowledge, such as the road and sand fortress that used the modest materials that KMB people could access, the simple house management of raising the floor and installing hanging storage, and the shelter options in the evacuation plan. Therefore, I argue that the embeddedness of this planning knowledge is generated and consciously shared through spontaneous and informal events, such as incidental meetings, *warung* talk, *arisan*, sidewalk conversation, and *kerja bakti*. These events allow the residents of KMB to perform reciprocal typification of what they have performed in past floods. The events are learning centers that allow them to share adaptive practices. They attend the events as routine activities without having to know the outcome in advance. The social events are the means by which they live in KMB, including attaining the knowledge of adaptation planning.

The structure of the *lifeworld* is a domain that increases the adaptive capacity of KMB people (Ehlert 2012; Hornidge and Schotes 2012; Oberkircher and Hornidge 2011). The poor do not and cannot depend on economic power as a means of adapting to floods. They rely on their own experiences. The KMB people have shown that they can use the flood to survive and even to earn an income. The repeated practices of how they choose escape routes, find shelters, build preventive facilities, and organize neighborhood cooperation are ways of increasing their adaptive capacity. These repeated practices have transformed the people of KMB mentally and physically to

5.5 Connecting to City Level: A Language of Planning Knowledge Converter Required

accept floods as ordinary. The poor are not vulnerable if they can apply the stock of knowledge in their community to use in creating local innovation.

However, the institutionalization of LEAP cannot be extended to the city level. KMB has specific traditions that are situated and embedded in the *lifeworld* of its people (Huxley 2000). A different *lifeworld* caused the disconnection between the *kampung* plan and *kota* (city) plan. As argued by Habermas (1987), mutual understanding can be achieved only if the representatives are ready to withdraw from power and start a rational argument. The disconnectedness indicates that locally embedded adaptation planning can be similarly associated with covert planning (Beard 2002). However, in this case, instead of being driven by social justice, it is caused by different realms and the actors' disconnected world. Moreover, KMB people still consider the government as part of their world and as potentially useful in providing facilities for adaptation. In other words, in contrast to Briassoulis's claim that informal planning is not necessarily institutionalized, I contend that LEAP can be combined with and institutionalized in the planning system, but it needs a converter that can translate the different emerging languages into a common language that all sides can understand.

Chapter 6
Planning and Adaptive Knowledge

Abstract The *lifeworld* of kampung people delineates their operational zone of planning and then produces a stock of knowledge for developing and managing their adaptation pathways. The interconnected meanings of vulnerability, adaptation, and planning constitute a trilogy of knowledge, which is generated from the lived experiences of *kampung* people when they reflect on their adaptation practices. The locally embedded adaptation planning (LEAP) contributes to the ongoing debate on adaptation and managing flood risk on three levels. First, the LEAP is an empirical evidence of planning process that practiced by *kampung* people. Second, it contributes to the discussion on how adaptation planning and effective response strategies built in the everyday life. Third, it offers practical recommendations for strengthening human dimension to current planning policies and practices.

Keywords Human • Knowledge • Adaptation • Flood risk

The lifeworld of kampung people delineates their operational zone of planning and then produces a stock of knowledge for developing and managing their adaptation pathways. The interconnected meanings of vulnerability, adaptation, and planning constitute a trilogy of knowledge, which is generated from the lived experiences of kampung people when they reflect on their adaptation practices. The locally embedded adaptation planning (LEAP) contributes to the ongoing debate on adaptation and managing flood risk on three levels. First, the LEAP is an empirical evidence of planning process that practiced by *kampung* people. Second, it contributes to the discussion on how adaptation planning and effective response strategies built in the everyday life. Third, it offers practical recommendations for strengthening human dimension to current planning policies and practices.

6.1 The Social World as a Domain of LEAP

This thesis sheds light on the use of *lifeworld* theory in conducting adaptation planning. Thus, using a people-centered perspective, it clarifies the relationships among vulnerability, adaptation, and planning. As described in Chap. 2, however, it is not clear how this approach could be applied to the adaptation planning process. In the discussion of climate change adaptation, the literature reflects the relational framework between adaptation and vulnerability. I have argued that the people-centered approach can differentiate the vulnerable from the adapters in locations that are vulnerable, whereas the other approach tends to categorize all residents in such locations as being vulnerable. Therefore, I focus on examining the adaptive capacities of people who have lived in vulnerable circumstances for a long time in order to understand the "real" state of vulnerability and the differences between the vulnerable and the adapters.

I was attracted to investigate the *lifeworld* of flood-affected people in KMB by the interesting phenomenon that floods are not a deterrent to migration into the KMB area. Through this investigation, I first found that not all of the poor who live in this flood-risk area are vulnerable and that not all of those who are vulnerable are helpless to the extent that residents living in the same building have different degrees of vulnerability. Second, the lived experience is a key factor in increasing the adaptive capacity of KMB people whether they are tenants or owners, males or females, octogenarians or schoolchildren, and socially connected or isolated. It is based on the self-reflective practices of KMB people during the series of floods. These lived experiences have resulted in significant differences between people who are vulnerable and those who adapt. Third, the *lifeworld* of KMB people has constructed a perception of flooding that differs from that of natural science. From the scientific perspective, the flood is a hydrometeorological disaster; big floods follow a 5-year cycle, and tidal floods have a 16.4-month cycle. Floods are caused by large amounts of rainfall and high spring tides, which the land cannot absorb. These conditions are exacerbated by improper land use, land subsidence, and poor drainage. In contrast, the people of KMB perceive floods as common and frequent events that inundate their homes and possessions. They have no idea of when a flood will occur or how bad it will be. They perceive floods as unpredictable. The *lifeworld* of KMB residents, who have lived with regular floods for years and who interact with other flood survivors, has changed the meaning of floods. It not only has influenced their definition of who is and who is not vulnerable but also has defined the state of vulnerability and the kind of adaptation pathway to be taken.

Based on the empirical findings of this research, the *lifeworld* analysis contributed to understanding the differences in identifying vulnerable people and defining an area at risk of flooding. As described in Chap. 6, the flood experience is the main factor that distinguishes the vulnerable and the adapters. In a similar vein, Bengtsson et al. (2007) and Kuruppu and Liverman (2011) argued that vulnerability depends on the personal lives and daily interactions of social groups and individuals and their "properties or stressors" (Cutter 1996, cited in Fekete 2010). Furthermore, the

6.1 The Social World as a Domain of LEAP

lifeworld analysis defined the spectrum from vulnerable to adapter, helping to identify the position of individuals in space and time (Table 4.3). The *lifeworld* analysis revealed the realm of KMB people in perceiving floods and knowing flood-related vulnerability, which confirmed that the human dimension is an essential factor in measuring the vulnerability of a community.

Because vulnerability assessment is a part of adaptation planning, *lifeworld* analysis should be applied to the planning process. Based on three case studies of adaptation planning in KMB, that is, scenario planning for evacuation, spatial planning for house management, and infrastructure planning for neighborhood facilities, the findings showed that the adaptation planning at the community level was reflected in the lived experiences of the affected people. The case of KMB demonstrated the use of the *lifeworld* in explaining the reasons for planning based on reflective practices. The *lifeworld* of these flood-affected people structured the zone of operation of their adaptation. Even though the operating zone of planning was characterized by short-term perspectives, a neighborhood scale, and a problem-solving orientation rather than long-term views, a city scale, and a visionary orientation, the KMB people have constructed their adaptation planning. Schön argued that planning could count on "the experiences (theory-in-use) that were embedded in the logic of the action" (cited in McDowell et al. 2007, 10).

Furthermore, the structure of the *lifeworld* of the KMB revealed that it continues to be useful source in producing planning knowledge. As described in Chap. 5, neighbors and predecessors must be reflected in defining subsequent or future adaptation practices. The planning approach of the KMB residents is based on their reflections on what they have done and how they have understood the historicity of floods in their area. LEAP of KMB does not use a participatory method but is a construction of "intersubjective meaning" (Schütz 1967; Schütz and Luckmann 1974), which is intentionally shared by and among KMB people through their "habitualized actions" (Berger and Luckmann 1967), such as *arisan, pengajian*, and *kerja bakti*. This approach showed that in this research, the adoption of Schütz's *lifeworld* elucidated the planning process on a personal level, which began with exploring the inner experiences of the planning process. *Lifeworld* analysis is a means of examining the institutionalization of adaptation planning in a social world because the *lifeworld* can provide the context and the domain of adaptation planning processes.

KMB residents shared their knowledge of adaptation planning by forming "a locally situated form of knowledge" (Antweiler 2004) and by not depicting the flood as the enemy. They made plans not to prevent or to cope with floods but instead to modify their houses and surroundings in order to mitigate their consequences. They realized that frequent floods and serial major floods have forced to shape their zones of operation in adaptation planning. Through the shared learning framed by the triple-loop learning cycle discussed in Chap. 5 (Fig. 5.9), KMB residents transformed routine coping practices into an increased adaptive capacity. From reacting in the first loop to corrected reframing actions in the second loop and modifying adaptation plans in the third loop, the residents of KMB learned to live with floods, which constituted their locally embedded adaptation plan.

Individual reflections on their past adaptation practices produced the meaning of the collectively shared adaptation plan. This process took place informally in casual conversations and get-togethers in KMB. Such events expedited the embeddedness of this knowledge in the realm of the KMB residents. LEAP is an authentic method for people who have no formal training in planning. Their self-taught planning knowledge reflects their practices.

Because they produced the knowledge themselves, it is more easily transmitted throughout the *kampung*. They acknowledged the plan by interpreting it based on reflections on their own experiences or those of others. They discussed the substance of the plan in order to arrive at a decision that affected all residents. Then they distributed assignments based on the recognition of the capacity and willingness of each resident. The verbal agreements to act on the plan were not written down because the parties preferred hearing to reading information. The plan was amorphous and flexible and did not include sanctions. People who did not follow the agreement were ridiculed and treated with scorn and contempt. Therefore, the social world of KMB is a domain of the LEAP.

6.2 LEAP is a New Insight for Planning Debate

Based on the exploration and investigation of the practices of KMB residents, the institutionalization of locally embedded adaptation planning can be divided into six steps (see Fig. 6.1). This model connects the *lifeworld*, reflective practices, and institutionalization in an integrated planning cycle. This planning model not only addresses the problems that have been defined by the local people but also increases their ownership of the planning product. This model also enriches the incorporation of the *lifeworld* and reflective practices into their planning practices.

The first step is to identify the adapter as an informal planner who can facilitate the planning process. The process should be based on the locally habitualized actions of the community in practicing social activities. The adapter should be an individual who has extensive experience with floods. A recommended method of identifying the adapters is the *lifeworld* analysis. In adaptation planning for floods, the adapter is someone who has adapted to floods. As discussed in Chap. 2, Schütz's *lifeworld* is suitable because it reveals the structure of the *lifeworld* of flood-affected people in a way that is more personal than Habermas's *lifeworld*, which emphasizes communication.

The second step is to compile the precedents of any actions that have been practiced by the local people. Whether the action has succeeded or failed is irrelevant. This collection is important in determining the kind of framing that was produced in performing that action. Framing is important for determining the kind of solutions that are presented as options. As described in Chap. 7, I recommend Schön's reflective practices because they investigate the relation of theory in use and espoused theory in planning. In locally embedded adaptation planning, because the actors have no special training in planning, their practical knowledge is revealed.

6.2 LEAP is a New Insight for Planning Debate

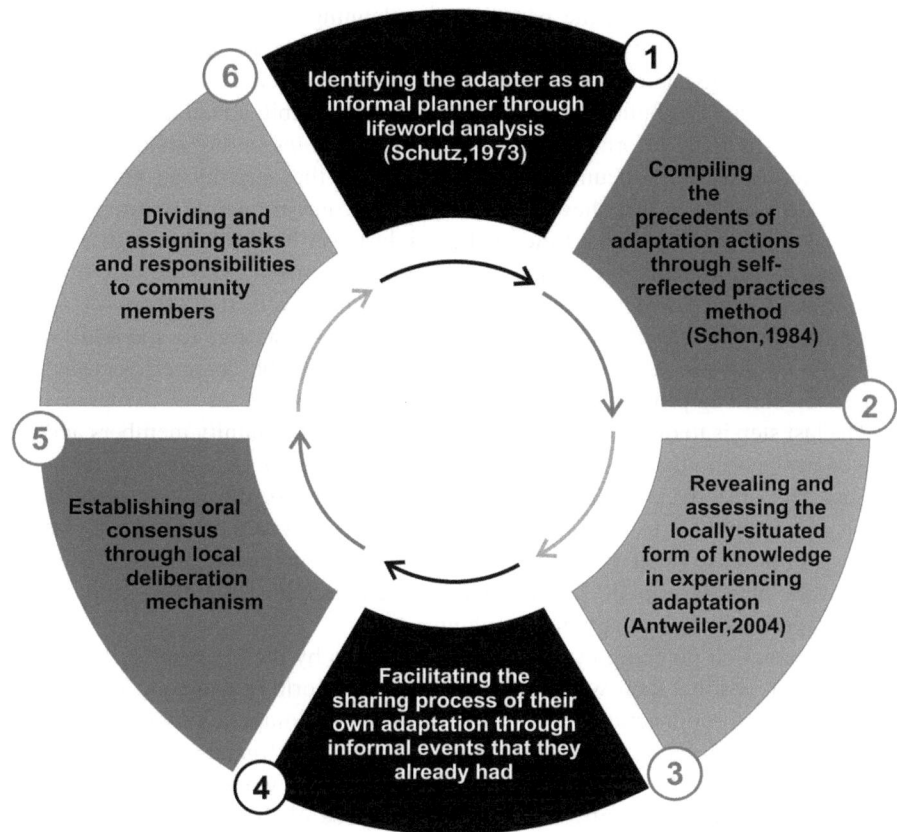

Fig. 6.1 LEAP process (compiled by the author)

The third step is to reveal and assess practical knowledge in the context of the sociology of knowledge. The conceptions of practical knowledge (e.g., experiential knowledge and locally situated knowledge) defined by Antweiler are useful in defining and categorizing the type of knowledge that is embedded in the realm of local people. In this research, the way in which KMB residents reflected on their adaptive practices and transmitted this knowledge to their neighbors and predecessors is a form of planning knowledge. The extent to which the projection resulted from the reflective practices depended on the structure of their *lifeworld*.

The fourth step is to facilitate a deliberative forum for discussing a social phenomenon in the community based on the traditions and habitualized actions of the local people. In this research, because the *kampung* is an informal settlement and the people have informal livelihoods, social events tend to be organized in a casual if not covert manner. Planning is imposed on those social activities without a clear framework or schedule. It is highly recommended to avoid any program that moves the participants from their own habitualized actions, such as the community aspiration meetings organized in the participatory planning approach. As discussed in Chap. 8, the advantage of using existing social events is that the participants can use their own language and communicate freely with each other.

The fifth step is to build a consensus based on the habitualized activities of the participants. In many informal settlements, such as *kampung*, the residents are not accustomed to reading documents or maps because they usually do not have the capacity to read them and they already know their environment. Improving their technical capacity would take time, and it might never be applied or institutionalized because both media—texts and maps—do not represent who they are or what they know. In the KMB context, oral consensus is an effective means of mainstreaming and institutionalizing the locally embedded knowledge for the KMB residents. By using the heads of *Rukun Tetangga* (RTs), the research focuses the consensus-building process on a single point of reference.

The last step is to divide the assignment among the community members and set up a control system that ensures that the plan will be implemented. In line with this process, the objectives of the plan would be met because the people understand floods and know how to live with them. The members would rehearse the plan in advance in order to make it as relevant as possible to their own interests. The oral form of the plan thus would be embedded in the community members' minds and then be applied in their adaptation planning.

The adaptation planning cycle that was generated by the practices of the KMB residents showed that their world is distant from the world of planners. The lack of resources and disconnection from formal institutions compelled them to concentrate on the world that is within their reach, which is the *kampung*. However, they still recognized that the governmental institutions in their world could help them to improve their adaptation infrastructure. The absence of the government or other formal institutions indicated that the locally embedded adaptation planning could be associated with covert planning (Beard 2002). However, in this case, I argue that instead of being driven by the need for social justice, locally embedded adaptation planning occurs because problems are perceived differently.

According to Gehl (2011), locally embedded adaptation planning uses an individual scale to examine problems. However, in planning, the practical knowledge of local people should be considered. In contrast to Gehl (2011), who suggested that planners and architects use the cognitive process in defining problems on a personal scale, this model uses *lifeworld* analysis to discover the lens used by common people who have experienced the problems. Therefore, based on the findings of this research, I call for a better humanistic planning model in answer to Friedmann's (2008) argument that planning needs to consider humanist philosophy. Planners should apply the LEAP model at the beginning of the planning project and subsequently strengthen the humanistic value in urban planning practices by using Gehl's personal scale.

6.3 LEAP is a People-Centered Adaptation

By developing this model of LEAP, this research contributes to the discourse of adaptation planning both in the climate change adaptation and disaster risk reduction discourse, which far has been dominated by the climate-proofing and community-based approaches. Even though LEAP does not fit to long-term aspect of climate change scenarios, it assures immediate local fit and applicability within the current context. These approaches have been predominant in both research and in practice, as discussed in Chap. 2. The model of LEAP provides an alternative to the predominant approaches to adaptation planning. It is worthwhile to discover the adaptation pathways of local people, especially those who have experienced a flood phenomenon.

This model also enriches the incorporation of the *lifeworld* and reflective practices into the ongoing debates of adaptation planning. As shown in Table 6.1, this model is more suitable than the other two models of adaptation planning for use at the community level. By depending on the lived experiences of local people and their reflective practices, this model encourages local people to take the initiative in making their community self-reliant. Therefore, I suggest that urban adaptation planning should be an accumulative process of the LEAP of urban communities.

This model should be linked to the city level because successful adaptation requires cooperation among urban stakeholders. Coordinated management that involves diverse stakeholder groups (Sperling and Szekely 2005) could assist in preventing maladaptation across sectors or on different scales as well as in creating

Table 6.1 Typology of adaptation planning

Elements of planning	Type 1 People centered	Type 2 Community based	Type 3 Climate proofing
Drivers/actors	Predecessors and people who experienced the climate-related disaster	Facilitators of CSO/NGO in CCA, DRR, and/or sustainable development	City planners or professionals who have a competence on CCA and/or DRR
Goals	Adapting to the changing situation	Increasing adaptive capacity	Reducing the level of vulnerability
Data	Practical situated knowledge	Climatic and non-climatic data/information	Climatic and non-climatic data/information and experts' justification
Methods/approaches	Intersubjectively reflective practice planning	Participatory planning	Rational comprehensive planning
Types	Problem-solving	Prioritized actions	Systematic response
Level	Individual to community level	Community to subdistrict level	District/city to national level
Source of knowledge	Lived experiences	Modules/guidance	Theories/practices

Source: The Author

solutions that could benefit urban stakeholders. The presence of LEAP as a product of vulnerable community—assumed by government—could contribute to effective adaptation planning at the city level. The model of LEAP could provide suitable adaptation measures that serve the most vulnerable groups, meet their most urgent needs, and increase their resilience.

IPCC (2012) stated that the knowledge of adaptation planning is still evolving and no single approach should be recommended. The key principle is that adaptation planning offers development benefits in the short term and reduces vulnerabilities in the long term. Instead of examining the role of donors and NGOs in urban adaptation planning as described by Carmin et al. (2012), I recommend that we should look at the planning practices in the community. Urban adaptation planning is more than the integration of technical and local knowledge, in which the latter merely supplements the former. It should also be used to disseminate the planning knowledge that has existed in many communities. Based on the adaptation planning in KMB, the urban adaptation planning process would probably shift the general planning procedure from a single process to one that integrates the LEAPs.

According to AKP (2013), several key issues of adaptation planning need to be addressed: (1) multiple layers of stakeholders, (2) ownership of the planning process, (3) money matters, and (4) shifting time horizons. As discussed in Chap. 4, NGOs that conduct community-planning projects in Jakarta, such as MCI, ACF, IAP, also face issues regarding the ownership of the planning process. The model of LEAP could address the ownership question because the plan is based on the *lifeworld* of the flood-affected people and has been institutionalized through informal events that are suited to and reified in their way of life. If the government connected this model to urban adaptation planning, the sense of ownership would be stronger than a program that had been created only by planning experts.

The adapters take the role of planners because they have already done some planning, such as assessing the flood situation, providing adaptation options, and organizing cooperation in the neighborhood. In addition to involving or even directing the planning process, the adapters also act as transmitters in order to institutionalize the planning products. Since they are a part of the community, it would not be difficult to use the local language as the main communication tool. Therefore, adaptation planning not only depends on the type of actor (Füssel 2007) but also on the lifeworld of the actor who is shaped by the intensity of lived experiences.

The triple-loop learning process demonstrated by the KMB residents also contributes to the discourse of integrating climate change adaptation into disaster risk management. In response to a specific disaster, the repeated coping strategy could be a source of adaptive capacity through the reflection process. It would shift the adaptation options of KMB residents in responding to the changing environment from reactive to anticipatory adaptation.

6.4 Contribution to Flood Management

As discussed in Chap. 3, the findings of this research showed that the divergent worlds of adaptation planning prevent institutions from delivering solutions to flood management. The different realms of planning result in inefficient and ineffective adaptations and might even create additional problems. Climate proofing cannot solve the problems at the community level, and locally embedded planning is not conducted at the city level. NGO involvement through participatory planning also has failed to bring the different realms of experts and vulnerable groups to a common understanding. The realms continue to exist on different levels instead of being structurally connected within a single, integrated frame.

Community-based disaster management is the most effective response strategy in flood management. It is the missing link between the disaster response that is actually needed and the response that is provided. Community involvement in planning and implementing the adaptation would lead to sustainable solutions. Therefore, in order to prepare a community for disasters, the role of the community in the planning process needs to be formulated.

Planning for flood management that does take into account the local shifting meanings of flood risk has precipitated the reliance on informal planning approaches by the poor in the *kampung* community. KMB residents well understand that their living space is an informal and flood-prone area. This knowledge encourages them to take a self-reflective approach to assessing their flood-related vulnerability. They voluntarily share the assignments for managing the flood plain area without formal regulation or guidance. They informally share their ideas on house protection. Finally, they do not depend on the kampung leader to organize flood management, but they keep him informed about the measures that they take. The people of KMB produce and reproduce these structures of flood risk management by using casual manner as their way of life.

The way in which the people of KMB live with the flood is incompatible with the government's flood management plan. The government has built the flood infrastructure to control the floods and reduce the flood risk. The zone of operation of KMB's residents is not wide enough to encompass the government's idea of flood control. They learn to live with the floods by gradually refining previous plans. The informality of flood management in KMB is acceptable because it is derived from the lived experiences of the residents. In contrast to the formalization strategy, I suggest that these informal institutions be fully recognized by being placed on par with and linked seamlessly to formal planning institutions.

Building resilience can also incorporate locally embedded planning at the community level where the process of planning is derived from the lived experiences of local people. Although the urban poor are always cast as victims of environmental changes (e.g., floods), they still have an intangible asset that allows the adaptation pathways that they have already planned to flourish. The KMB case study demonstrated that the residents developed their plans for the short term, and they formulate goals that are more practical than the goal of procedural planning. This process occurs because their thinking and actions are spatially structured by the kampung area, socially constructed by the informal system, and temporally limited by their predecessors' experiences. This *lifeworld* explicates their reachable world, which needs to be recognized before it can be connected to the bigger, different world of the city.

The institutionalization of LEAP would be useful in disaster risk management at the community level. This research showed that KMB residents have engaged in flood risk management by using their local knowledge. According to UN DESA (2003), flood risk management consists of risk assessments, including potential flood hazards and vulnerability, in addition to flood plain management, which is assessed by structural and nonstructural measurements. Similar to UN DESA (2003), based on reflective practices, KMB residents have practiced (1) understanding flood hazards based on the past experiences, (2) assessing flood-related vulnerability based on intersubjectively shared meaning, (3) planning house management and neighborhood infrastructures, and (4) managing evacuation from flooded areas.

Therefore, I conclude that flood management will remain ineffective if the planning knowledge of flood-affected people is not revealed and consolidated into official flood management practices. Flood management planning must take advantage of the practical knowledge of adaptation planning that is embedded in each community. The shifting focus in the adaptation planning process, from the merely climate-proofing approach to the locally embedded knowledge of vulnerable people, will connect the macro and micro perspectives of flood management. I argue that effective adaptation to floods cannot be institutionalized if there is only a single process of adaptation planning and if only experts are involved without considering the views of the real experts, that is, the *kampung* residents.

6.5 Implications for Urban Policies and Planning Practices

Many KMB residents have succeeded in becoming adapters who have no interest in moving; some have even managed to benefit financially from the floods. They perceive the floods not as crises but as normal occurrences. When a flood occurs, they modify their houses and surroundings. They use trial and error to actualize their knowledge and use reflected practice in planning adaptation. The lessons learned from previous floods assist them in preparing for the next one through the refinement of their plans. They have demonstrated that they have planning knowledge in their own language; thus, they customize traditional social events, including *kerja bakti*, in order to transmit planning ideas to other community members. Therefore, the planning practices in KMB are informally organized, and they produce several informal planners based on their zones of operation. These informal planners play a significant role in building community resilience, especially by sharing lessons learned with colleagues and family members, by advocating evacuation and shelter plans in the neighborhood scale, and by managing the flood protection facilities at the kampung level. Therefore, they should be included in urban resilience planning.

In KMB, *kerja bakti* extends beyond intergroup cooperation among residents used to maintain social cohesion (Huber et al. 2004), "traditional institutions" (Wilhelm 2011), "collective activities" (Lont 2005, 42), or "duty work" (Perkasa and Hendytio 2003, 130). It is also a medium for the exchange of knowledge. The regularity of *kerja bakti* ensures the reciprocal knowledge sharing and thus structures the *lifeworld* of the individuals and institutions within the *kampung*. Therefore, the city government should understand and consider this phenomenon planning urban development.

According to Joko Widodo, a former governor of DKI Jakarta Province and president of the Republic of Indonesia, Jakarta's development should focus on improving the *kampung*. The first step is to understand the *lifeworld* of the *kampung* people. The next steps are to discover the locally embedded planning that has been institutionalized in the *kampung* and then to connect it to the city's planning institutions. This research has shown that KMB residents have used reflective practices to plan their adaptation pathways according to their lived experiences on the knowledge shared by their predecessors and neighbors. Even though this research is limited to the single case of the KMB, it could be used as a stepping-stone to further research on the applicability of this model of LEAP to other *kampung* and/or gated communities in the cities and the rural communities of Indonesia.

The findings of the research suggest that the urban development planning system in Indonesia should provide an enabling environment for this type of planning. LEAP could be integrated with city planning through following several recommendations. First, the city government needs to evaluate the procedure of the annual *musrenbang* (planning forum), which is held at various levels from the *kampung* to the city level. At the *kampung* level, the government should provide more time and open the floor to informal planners to disclose their locally embedded planning. Second, the cities need to discover a "plugged-in" application that would compile the locally embedded planning of communities. This "plugged-in software" would also enable the cities to extend the application of planning, to facilitate the addition new features in the planning process, and to reduce the uniform codification of planning. Third, the central government should review the planning regulations[1] to support this plugged-in software in order to enable its use by different cities. This customization is imperative if the localities are to be included in the development of

---

[1] Planning regulations in Indonesia are distributed among several ministries. In this case, at least four regulations are needed to create this plugged-in system: (1) *Surat Edaran Bersama Menteri Negara Perencanaan Pembangunan Nasional/Kepala Bappenas dan Menteri Dalam Negeri* Number 0008/M.PPN/01/2007 and 050/264A/SJ, 2007 regarding technical guidance of *Musrenbang*; (2) *Peraturan Kepala Badan Nasional Penanggulangan Bencana (Perka BNPB)* Number 4, 2008 regarding technical guidance of disaster management planning; (3) *Peraturan Menteri Pekerjaan Umum* Number 20, 2011 regarding the guidance of detailed spatial planning; and (4) Peraturan Menteri Dalam Negeri (Permendagri) Number 66, 2007 regarding rural development.

the cities. An interface model that suits the purpose or tastes of different urban communities would facilitate this collaboration.

With regard to planning practitioners, the findings of this research suggest that the community of planners needs to learn about the model of locally embedded adaptation planning because it can help them to address Baum's (1980) critiques, in which he argued that planning practices are not always connected to the real world. Furthermore, adaptation involves human-environment interaction that should be examined as a social phenomenon. This phenomenon includes vulnerable people who experience floods and want to become adapters by learning from their experiences, their predecessors, and their surroundings. Triple-loop learning has transformed their adaptation pathways from being reactive to becoming anticipatory through reflective practices. Therefore, planners need to understand the *lifeworld* of the people with whom they interact through the model of LEAP.

This research focuses only on the phenomenon of the informal settlement, which is likely irrelevant in formal urban settlements, such as gated communities and apartment complexes. Moreover, it does not reflect the characteristics of rural communities. The KMB is a plural community at an economic level that is similar to a rural community. However, the model of locally embedded adaptation planning has not yet been tested in mixed economic communities. Therefore, I recommend testing the model of LEAP in different types of communities. Subsequently, it would be necessary to conduct additional research to develop a converter tool for incorporating LEAPs in the planning cycle at the city level.

Appendices

Appendix 1: Methodology

Research Design

Because this is an argumentative-exploratory study, it combines quantitative and qualitative methods (Creswell 1994). The quantitative method will be used to select a *kampung* that best represents the interplay between the poor and the floods described in the previous section. In this chapter, I describe the preparation of the qualitative research design, which in this case is a phenomenological study. According to Creswell, "a phenomenological study describes the meaning of the lived experiences for several individuals about a concept or the phenomenon" (Creswell 1998, 51). Thus, this combined approach will be used to explain adaptation planning and present a detailed view of the perception, meaning, and institutionalization of locally embedded adaptation planning through the example of the *kampung* in Jakarta.

This research is conducted in a natural setting without the application of controls to the situation of respective *kampungs*, which, according to Patton (1980, 42), contrasts "experimental research where the investigator attempts to completely control the condition of study". The natural setting provides a holistic and in-depth perspective and a unique case orientation. I analyze the data inductively without a static hypothesis by considering continuously evolving facts that have context sensitivity until I find the answers to the three research questions. As the researcher, I maintain a nonjudgmental bias in observing and describing group patterns, similarities, and differences.

Phenomenological research explores a shared experience among people who experience the same phenomenon in order to know its meaning. What is experienced and how people experience it are the main research questions that can lead to understanding the "real meaning." Mishler (2000, 10) argued, "Meaning should be perceived within the social context in which it occurs." Therefore, in this study, the challenge is to minimize multi-interpretations by either myself, the researcher, or by the people in the *kampung*, who were the participants in the study.

I base this phenomenological research on Schütz's theory of the *lifeworld* to understand how people in *Kampung Muara Baru* (KMB) perceive and respond to floods. The *lifeworld* comprises experiences shared by individuals through their own knowledge and actions in adapting to stress or shock caused by floods. The *lifeworld* can explain both individual experience and the universal nature of floods in the *kampung*. Thus, the present study is a narrative that tells not only an individual's life story but also the meaning of the lived experiences of KMB's people.

Field Work Procedure

Before conducting the fieldwork, I prepared to make it as easy as possible. I set up my desk in my department in the *University of Indonesia*, recruited research assistants, and contacted the government of DKI Jakarta about the fieldwork. I conducted the fieldwork in KMB from mid-April 2012 to the end of March 2013. I collected data and information concomitantly with the secondary data collection in several local and national government institutions and related agencies. I received a small grant for the fieldwork under the joint cooperation of the International Secretariat of Global Change System for Analysis, Research, and Training (Table A.1).

Table A.1 Fieldwork schedule

No	Periods	Activities
1	01.04.2012–30.04.2012	Preparation: going to Jakarta, setting up host in urban studies postgraduate program UI, recruiting assistants, gaining research permission from DKI Jakarta Province; providing and testing list of questions in Indonesian terms
2	01.05.2012–22.06.2012	Secondary data gathering, preparing maps; interviewing head of RW and planners
		Conducting mental mapping activities and participant observation for first and second research question
3	22.06.2012–16.07.2012	Field observation; mental maps cross-checking; informal interview; key informant discussion; and group interviews for second research questions
4	16.07.2012–16.08.2012	Participant observation excluding the political campaign, preparation, and election days (5–7 July)
5	16.08.2012–30.09.2012	Making data transcription; translating into English
6	01.10.2012–10.12.2012	Focus group discussions in three different RWs; storytelling and case study; in-depth interview for third questions
7	12.12.2012–12.03.2013	Making data transcription and translating into English and conducting participant observation, if flood occurred, for cross-checking FGDs' result

Before arriving in Jakarta, I designed my fieldwork plan by considering the factors that could influence the data collection, especially community meetings. The first factor was the election of the governor of DKI Jakarta, which prohibited meetings during campaigns and on election days. The second factor was the fasting month and *Idul Fitri* (Eid Mubarak day), which were not the best times to talk to people about floods. The third factor was the prediction of big floods in late 2012 and in early 2013, which could be suitable for participant observations. In the field, most plans worked well, except some unexpected adjustments necessitated by the second round of the election and the fasting month in September 2012. However, these changes did not affect the overall plan.

I recruited two research assistants (one male and one female) for 4 months to help with collecting basic data and conducting workshops and meetings. Both were university students with a background in geography. In order to collect good data, I always briefed them before we visited the *kampung* area and engaged with the *kampung* dwellers. I emphasized the importance of exploring and recording the local language during interviews and group discussions.

To support the data collection, I used my office in my home institution's postgraduate program in urban studies as the base where I organized the fieldwork. I also rented a one-room apartment in the KMB neighborhood to support the participant observations. With the interviewees' permission, I used a voice recorder to record the interviews, a laptop-camera to shoot the group discussion activities if it was convenient, and a camera to take some photos. This logistical support was provided by grants from START and IAP.

I started my fieldwork by introducing myself to the head and secretary of KMB, explaining my activities, and asking him for a list of the head of *Rukun Tetangga* (RT) (smaller units constituting a *Rukun Warga* (RW)). I also interviewed them separately. My assistants and I then verified the maps and statistical data related to KMB by conducting field observations there. We conducted semi-structured interviews with the heads of the RTs, asking them about their area, population, flood history, and about the people who had the most and the least experience of floods.

I used this information to choose three neighborhood units that represented the flooded area but were in different geographical locations in order to explore the lived experiences of flood-affected people. I selected them based on my judgment and the aim of this research as suggested by Greg and Taylor (1999), Schwandt (1997), cited in Groenewald (2004). I looked for people who had experiences relating to the flood phenomenon. The first was RT 21, which is next to Jakarta Bay, the fisheries harbor, and the water pump. The second was RT 7, which is in the middle and surrounded by factories and warehouses. The third was RT 15, which is next to Sunda Kelapa harbor. I began a snowball survey, starting with the person who had been recommended by the head of RT. Snowballing is a method used to expand the sample by asking one informant or participant to recommend others (Babbie 1995). When I found someone who had a good memory, I followed him/her up by tracing the oral history or conducting an in-depth interview.

In each RT, I conducted several group meetings on different topics and with different aims. I elicited their perceptions of the floods and the extent of their flood losses and damage. I then asked how they had coped with and adapted to the flood situation, especially in relation to themselves, their families, and their houses. I provided RT maps to assist them in identifying shelter points and evacuation routes. I did not help them to draw on the map but let them visualize their experiences on the map. Next, I asked them how they came together to help each other find the neighborhood adaptation pathway.

Next, I interviewed some neighbors who had not attended the group discussion, including shopkeepers and factory and warehouse workers. After they became comfortable with me and I had obtained the information that I needed, I returned several times when I need to clarify unclear or incomplete information. I observed the participants in each RT. I was there when KMB was flooded in January 2013. For 2 weeks, I observed the situation and how the locals interacted with outsiders, such as the *kelurahan* (the smallest government unit) officers, the police, and business owners (warehouses and the fisheries port) while they adapted to the flood. I gathered the information that I had needed to confirm their stories. It was an invaluable experience, and it generated rich data for my research.

I also had conversations when I met people in the alleys of *warungs* (small shops). I attended social meetings, such as *Rembuk RW*, *Arisan* off-talk, and *pengajian* (religious group discussions). The *Rembuk RW* is an annual community meeting with the government to identify problems and needs and solve problems based on the priorities of the *kampung*. The *Arisan* is a form of rotating savings association that meets once a month. It can be based on gender, social group, or hobby. In KMB, most *arisans* are for housewives. The arisan often coincides with *pengajian* (Quran reading). Therefore, in KMB, the *arisan* is motivated by economic need and social and religious interests.

An alley chat is a casual conversation between neighbors. Because the alleys are only 1.2–1.5 m wide, it is very easy to talk to the person living opposite. Men usually have these conversations on their way to work in the morning or in the evening after dinner. Women have these chats on their way home from the market in the morning or in the afternoon when they bring in their laundry. The neighbors talk about activities in their daily lives and about managing their living places. It is also an effective means of sharing news because they stand in front of their houses and can directly observe the new adapted form.

Warung talk is a casual conversation in the *kampung*. A *warung* is a small shop attached to a house where necessities, such as soap, oil, medicine, book, candies, and even cigarettes are sold. In the neighborhood, there are seldom *warungs* that serve only food, but some *warungs* in the kampung are an exception. Both men and women visit the *warung* on a daily basis and it is a meeting point. A *warung* conversation about the latest news can be informative, and these conversations are open to everyone.

Under the Indonesian Association of Planner (IAP), this research considered the issue of confidentiality and the informed consent of respondents. I told my interviewees that their responses were being collected for academic reasons. I also explained to them that the data would be confidential, and I asked for their permission to use their names if the research was published. Some of them gave their approval, such as the Izhar Chaidir (government), Gustara (the head of KMB), Konedy (Secretary of KMB), and some heads of RTs. However, informants RT 19, RT 21, and RT 15 did not give their consent. I promised anonymity to those who did not want their names to be used. In many cases, the interviewees, such as the Secretary of RW 17 and the heads of RTs were not comfortable being recorded or photographed. With their permission, I took written notes of my interviews with them.

My assistants and I informed the respondents about the purpose and the outline of the research. We then asked for their informed consent to conduct interviews, group discussions, and take oral histories. Consent was given in writing or orally (Appendix 2). I also promised to give them a summary of the research upon request at the completion of the study.

Steps in the Data Processing

I stored the data in the forms of audio and video recordings and field notes. I recorded interviews only with the interviewee's permission. I labelled each interview with codes, such as "participant, 13 July 2012." For participants who were interviewed more than once, I added a number on the code, such as "Participant-1, 13 July 2012." I made notes and transcribed keywords, sentences, and phrases to document the original language of the KMB people. In addition, I stored the data in the field notes to avoid losing or forgetting important information.

Field notes are important in recalling data in qualitative research (Lofland and Lofland 1999).

I had kept types of field notes. The observational notes recorded my activities when I observed KMB and interviewed people if they did not want to be audio recorded. I also summarized my findings. The other type was a methodological note that recorded self-instructions or briefs for the assistants before and after gathering data. Both field notes included my interpretations but not categorizations. I wrote those field notes to clarify the data from each interview (Groenewald 2004; Caeli 2001).

I then categorized the data based on my main research questions. I did three rounds of data processing: conceptualizing flood-related vulnerability, process tracing to construct locally embedded adaptation planning, and analyzing the *lifeworld* to explain the institutionalization process.

Defining vulnerable people was the key to beginning the research. Based on previous research that identified the Penjaringan subdistrict as a most vulnerable area, its residents were selected as the most vulnerable people. Based on the definition of vulnerability discussed in the previous chapter, data were collected in Penjaringan, including slum and squatter settlements, demographic profiles, and climatic and physiographic conditions related to the flooded area. These quantitative data were gathered from the secondary data of the Jakarta Provincial

Government, particularly the spatial planning division and statistical office. The data included social demography, land use and housing, livelihoods, and basic settlement infrastructure.

The spatial data were digitized and projected on the geo-reference maps and analyzed to determine the vulnerable areas. At this stage, the effects of climate change that were assessed were the rise in sea level, high tides, heavy rains, and river overflows that flooded the KMB. In order to confirm the information, qualitative data were gathered at the subdistrict level through newspapers, recent studies, and government reports and at the RT/RW level through historical transect mapping and semi-structured interviews with the heads of RTs, the government officers of DKI Jakarta (head of the subdistricts and head of the planning division), NGOs with projects in DKI Jakarta (Mercy Corps, Action Contre La Faim, and the World Bank), and professional consultants who had conducted planning in Penjaringan. The historical transect focused on the chronology of floods, the frequency and locations of disasters, and the short- and medium-term adaptation measures taken by the poor. Combined, this information produced a map of vulnerability that later became the base map for conducting transect walks with the heads of RTs and RWs for the clarification and verification of the affected houses. The interviews with the heads of RTs were conducted to confirm the history, the time, the area, and the impact of floods. Based on these data, the internal and external perceptions of the flood area were juxtaposed.

In addition to the flood map and history, I interpreted the interviewees' statements and opinions with attention to clarification and elaboration, and I asked for facts, feelings and perceptions (Mielke 2011) of institutions, planning processes, involvement of vulnerable people, community profiles, and best experiences. All qualitative data were assessed using interpretative analysis (hermeneutics) in order to understand what they intended to have meaning with symbols on their own maps and cultural expressions in their texts and the connections among them (Bernard 2006). The results of this analysis provided a general picture of the umbrella institutions of KMB and the discourses of adaptation planning.

In the next step, I profiled the identified RTs and discussed the flood experiences with the heads of these RTs. Through RT meetings (group interview), I elaborated households' perceptions of vulnerable people and described the RTs as the focus area in the third phase of analysis. In these RTs, I elaborated the interviews based on the recommendations of the heads of the RTs, who were the *gatekeepers* of their neighborhoods. According to Neuman's (2000, 352) definition, a gatekeeper is someone with the formal or informal authority to control access to a site.

I then defined the adaptation planning. Using information obtained from the gatekeepers, I explored the data collected from the flood-affected people. In addition to the in-depth interviews, I conducted social and resource mapping and group discussions to determine the planned adaptation that had already been conducted, was being conducted, or would be conducted based on their experiences, knowledge, and perception. The social mapping focused on the location of the vulnerable groups, their shelters, and their planned escape routes. In addition, the resource mapping focused on the distribution of public facilities and their assets. Both were used to explore the spatial and temporal dimensions of the people's reality (Kumar 2002).

I conducted *participation observations* in order to learn more about the community and its daily activities. I observed and participated in a small-scale social setting, usually in the researcher's home culture (Neuman 2006). In this stage, the researcher becomes an observer participant because he/she needs to be closer to the people who have covertly designed an adaptation plan or have participated overtly in formal planning. I wanted to learn about their cultural backgrounds, stories, and variations of life, components of action, and exceptions (Dewalt and Dewalt 2002).

Quantitative data were needed to help the mapping process of both the *kampung* selection and group discussions in the selected *kampungs*. In the fieldwork, maps provide initial information about roads, coastal lines, and rivers, which could be used later by the vulnerable people to compare with their own understanding of the places, various adaptations, and their planning. This spatial information revealed the shared symbols that the informants might use in mapping adaptation plans. During the mapping, I observed their way of communication, the local words used in dialogue, and the agreed-upon symbols.

The qualitative data were gathered through in-depth interviews, participant observations, and field notes. The interviews were conducted privately in a confidential atmosphere. Field notes were used to record dialogues or events that were relevant to the study. These included a compilation of maps, photos, audios, other handwritten documents, and an inventory file, which were consolidated later to organize the data, track the progress of the fieldwork, and communicate with my supervisor. The data were explicated using a phenomenological methodology based on Schütz's *lifeworld*. According to Vanderberg, Schütz said, "the human world comprises various provinces of meaning" (cited in Groenewald 2004, 4). Therefore, an iterative process is used to deepen the understanding of the *lifeworld*. This process was useful in producing meanings.

In the next step, I constructed the institutionalization of adaptation planning. The hardest part was to recognize the participants' activities in relation to planning. Having identified the people who had experienced planned adaptation, I observed them in meetings and group activities in the RT/RW/*kampung* to learn how they communicated with their friends in the community. In my field notes, I listed the people who had experienced floods and any group activities that related to their adaptation. In addition, I conducted in-depth interviews with people who had been affected by, observed, or remembered past floods. I used storytelling, which is a way of organizing information, conveying emotions, and building community (Sturm, 2007), to interview the elders and/or community leaders who knew the history of the *kampung*, including the norms, the influencing actors, and community groups that had planned adaptation. In addition, I conducted serial focus group discussions (FGD) in order to understand the institutionalization processes of planned adaptation in the community at the present time, particularly the rules, actors, and (authoritative) resources. I assumed that the rules (procedures that people follow in their social lives) were produced and reproduced by KMB dwellers in their practices. The FGD included eight to ten participants who had made a practice of making rules or providing resources and access, especially in relation to floods. I gathered data on opinions, attitudes, and discussions about adaptation planning. I examined the FGD data using the narrative analysis of framing to discover regularities in how people told stories about adaptation planning. Framing is utilized to elaborate the social process phenomenon, which is a part of the construction of meaning by participants and opponents (Snow and Benford 1988).

The last phase of data processing involved the explicitation of the data. According to Hycner (1999), this phase consists of five steps: "(i) bracketing and phenomenological reduction, (ii) delineating units of meaning, (iii) clustering of units of meaning to form themes, (iv) summarizing each interview, validating it, and where necessary modifying it, and (v) extracting general and unique themes from all the interviews and making a composite summary" (Groenewald 2004, 17).

In bracketing and phenomenological reduction, I separated my personal views on the flood phenomenon from those of the KMB dwellers. I avoided including preconceptions in my interpretation. Phenomenological reduction is a deliberate and purposeful opening by the researcher of the meaning of a phenomenon in its own right. Therefore, I had to familiarize myself with the words of the interviewees by listening to the audio records several times and/or reading their words recorded in my field notes. I paid the most attention to the unique experiences of the KMB people in adapting to floods.

Similar to Hycner (1999), I used bracketing to delineate units of meaning. I considered the verbatim content, the number of times that a meaning was mentioned, and how it was stated (verbally or nonverbally). The same unit of meaning could have different weights or chronologies of events. I explicated the data by carefully extracting each interview or discussion of KMB's dwellers and consciously bracketing my presuppositions. Next, I examined the list of nonredundant units of meaning. To avoid overlapping clusters, I defined central themes through the cross-examination of the meaning of the clusters. The central themes "express the essence of these clusters" (Hycner 1999, 153).

The following steps summarize the units of meaning by incorporating all the themes extracted from the data, validating them by returning to the interviewees' words to determine whether the essence of the interview had been correctly captured, and modifying it if necessary. These steps depend on the validity check process. In the final step, I generalized the themes that were "common to most or all the interviews as well as individual variations" (Hycner 1999, 154). I wrote a composite summary that reflects the context or "horizon" from which the themes emerged to conclude the explicitation process. The summary is a narrative of how vulnerable people institutionalize adaptation planning to develop the "essential structure" of their social world. This conclusion is followed by the implications of the findings for both science and policy-making.

Research Constraints

I encountered several difficulties in conducting this research, especially during the fieldwork. These included time-survey constraints, sociopolitical conditions, and problems in processing the data. As previously mentioned, my fieldwork was interrupted by the gubernatorial election. Because it lasted two rounds, there were about 30 days when I could not interview anyone. During this time, if the KMB people had met with an outsider, they would have assumed that he/she was campaigning for one of the candidates. Because the floods have been seen as a political issue, their answers would reflect "the truth" of their opinions. The supporters of the incumbent tended to be defensive in their answers. They often said that the flood infrastructure had reduced the frequency of floods. The opponents of the government were more likely to complain. In order to overcome the time constraints, I reviewed my interviews (see Appendix 3) before and while gathering the secondary data (see Table A.2) from the government office. With regard to the sociopolitical constraints, I emphasized at the beginning of interviews, or during the interviews, that my focus was on learning about their actions, opinions, and resources in adapting to floods. I shifted the direction of the discussion if he/she was for or against the government. However, after the new governor was elected in October 2012, I returned and tried to elicit their opinions about the government programs.

In the data processing, it was difficult to bracket my personal view of KMB because I had lived in a *kampung*. On one hand, I had the advantage of understanding it, but on the other hand, my interpretation might have been biased. In my interviews with the KMB dwellers, I tried to avoid links or words that that I had used in my own *kampung*, even the same words. Thus, I kept asking them to elaborate. Even though absolute objectivity was impossible, I was always aware of the potential biases in my research, and I tried to compensate for them.

Table A.2 List of secondary data

No	Items	Institution
1. Spatial data		
1	Basic map of North Jakarta	Bappeda DKI Jakarta
2	Land use map 1:5000	Dinas Tata Ruang DKI Jakarta
3	Spatial plan of Jakarta (RTRW)	Bappeda DKI Jakarta
4	Detailed spatial plan of subdistrict of Penjaringan	Dinas Tata Ruang DKI Jakarta
3	Vulnerability map of Jakarta due to sea level rise (based on USGS methods)	BRKP
4	Socioeconomic map of Jakarta	University of Indonesia
5	Flood simulation map	Institute Technology Bandung; World Bank Jakarta; University of Indonesia
6	Land subsidence map	Institute Technology Bandung
7	Risk map of Jakarta	BNPB

(continued)

Table A.2 (continued)

No	Items	Institution
2. Statistical data		
1	Penjaringan subdistrict in numbers 2011	BPS DKI Jakarta
2	Number of slums and squatter area in Penjaringan subdistricts	Dinas Perumahan Jakarta
3	Number of poor household in Penjaringan subdistricts	Dinas Sosial DKI Jakarta
3. Reports		
1	Final report, spatial planning RTRW 2011–2030	Bappeda DKI Jakarta
2	Final report, detailed spatial planning of Penjaringan subdistrict 2012–2032	Dinas Perumahan Jakarta
3	Final report, risk map and disaster management plan 2011	BNPB
4	Final report policy dialogue of North Jakarta case study 2011	IAP
5	Final report, urban poverty and climate change in Jakarta 2010	World Bank
6	Final report, Jakarta coastal development strategy 2010–2011	Bappenas
7	Final report, strategic environmental assessment of North Jakarta 2010	Bappeda DKI Jakarta

Appendix 2: Participant Information

Research Title
Climate Adaptation Planning: The perception, meaning, and institutionalization of locally embedded adaptation planning along the example of Kampung Penjaringan, Jakarta, Indonesia

Contact
Hendricus Andy Simarmata
s5hesima@uni-bonn.de
simarmata.andy@yahoo.com

Introduction

This research is a PhD research based in the ZEF-Center for Development Research, University of Bonn, to be carried out in DKI Jakarta, Indonesia. The research will be carried out from 28 March 2012 until 28 March 2013. I hereby request your assistance as respondent in this research because your knowledge and experiences are useful for constructing the conception of institutionalization processes of locally embedded adaptation planning. The purpose of this research is to understand the institutionalization of climate adaptation planning from individual lifeworld's perspectives in the local community through identifying the vulnerable people who has practices of planned adaptation, unfolding the meaning of adaptation planning based on their perception, and understanding the way of planning process in the community.

Participation

Your participation in this project is voluntary yet very much appreciated. If you do agree to participate, you can withdraw from participation at any time without comment or penalty.

You can participate in this research by:

1. Participating in interview discussing your perception on vulnerability, adaptation, planning, and institutionalization process, which are understood and practiced by yourself
2. Following in a group discussion to exploring the idea of vulnerable people, meaning of adaptation planning, and what kinds of rules, actors, and resources that you think that you and your community had
3. Involving in social mapping to defining the history of vulnerable places, adaptation measure taken, and the symbols that used by your community

Expected Benefits

It is expected that this research will benefit the policy-makers in Jakarta and other cities in Indonesia in the field of urban planning and climate change adaptation; the actors, including communities; government of DKI Jakarta; and professional association, as it will produce an explanation of institutionalization process of adaptation planning, which may you and your community had been done. At the end, this research may result in responsive knowledge sharing/exchange of climate adaptation planning, especially from developing countries.

Confidentiality

The names of individual persons will not be used, and the names of institutions will be used only with consent. I will use the information for my academic written only, and if I propose to use any quotes recorded during an interview or workshop, we will verify with the respondent and/or participant that the quote is truthful before its inclusion.

I intend to make audio recordings of interviews and transfer it into a transcript for the analysis works in the research. You may refuse to be recorded and still participate in this research. If you refuse to be audio recorded, I will take notes to record the interview. Audiotapes will be destroyed 5 years after the research has been completed. The records will be kept privately and transcripts will be stored in password-protected files. Audio recordings will not be used for any other purposes. Access to audio recordings will be restricted to the assistant researcher, researcher, and supervisors of this research.

Consent to Participation

I would like to ask you and/or your institution to confirm verbally your agreement to participate.

Questions

Please contact me in the above email address if you have any questions answered or if you require further information about the research.
  Thank you.

Appendix 3: Semi-structured Interview Guidelines

Planners

This guideline will provide semi-structured questions for an interview for planners. This research will investigate mainly the process of adaptation planning, the roles of planners, peoples' participation, and the institutions. The questions in here however will be aimed at asking the planners' perception on the role of people when conducting the planning process.

Adaptation Planning Process

1. What is adaptation planning in your perception?
2. In which level(s) do you think it should be conducted? Please explain.
3. What kinds of approach that you preferably choose to conduct?
4. How many times do you conduct the planning process, which related to climate change/disaster?
5. What do you think about the validity of data and climate model?
6. Did you feel satisfy with the process? Why?

Role of Planners

1. Where do you learn about adaptation planning? How do you perceive it?
2. What are your tasks during the process?
3. Can you explain about your function in the process?
4. In which term do you think you need to increase you knowledge or skills?

People's Participation

1. How do you explain to people about adaptation planning?
2. According to you, did they understand well?
3. What do you think about level participation of people in your case?
4. What are the useful things that you can get from the people? Knowledge? Experiences?
5. According to you, is it a must that you should consider people's knowledge? People's experiences? Would you mind to elaborate it more?

Institutions

1. Which rules/regulations become your references in conducting adaptation planning? Did it work?
2. Do you think we need a new regulation to conduct adaptation planning?
3. In the community level, do you think it is necessary to build an organization to plan and implement the adaptation plan?
4. Who do you think are the most important actors in adaptation planning?
5. What kind of resources that can be used at the community level?
6. How do you see the relation between community and government in synchronizing the adaptation plan?

   Thank you.

NGOs

This guideline will provide semi-structured questions for a leader of NGOs. This research will investigate mainly the process of adaptation planning, the roles of NGOs, peoples' participation, and the institutions. The questions in here however will be aimed at asking the NGOs' perception on the role of people when conducting the planning process.

Adaptation Planning Process

1. What is adaptation planning in your perception?
2. In which level(s) do you think it should be conducted? Please explain.
3. What kinds of methods that you preferably choose to conduct?
4. How many times do you conduct the planning process, which related to climate change/disaster?

5. What do you think about the validity of data and climate model?
6. Did you feel satisfy with the process? Why?

Role of NGOs

1. Where do adaptation planning projects come? Have you learned it before?
2. Do you have referenced organization? Would you mind to tell story about it?
3. Can you explain about your (institutional) function in the process?
4. In which term do you think you need to increase you knowledge or skills?

People's Participation

1. How do you explain to people about adaptation planning?
2. According to you, did they understand well?
3. What do you think about level participation of people in your case?
4. What are the useful things that you can get from the people? Knowledge? Experiences?
5. According to you, is it a must that you should consider people's knowledge? People's experiences? Would you mind to elaborate it more?

Institutions

1. Do you refer to rules/regulations or international best practices when conducting adaptation planning? Does it work?
2. Do you think we need new rules or regulation to conduct adaptation planning?
3. In the community level, do you think it is necessary to build an organization to plan and implement the adaptation plan?
4. Who do you think are the most important actors in adaptation planning?
5. What kind of resources that can be used at the community level?
6. How do you see the relation between community and government in synchronizing the adaptation plan?

Thank you.

Government Officers

This guideline will provide semi-structured questions for government officers who have authority related to adaptation planning. This research will investigate mainly the process of adaptation planning, peoples' participation, and the institutions. The questions in here however will be aimed at asking the NGOs' perception on the role of people when conducting the planning process.

Adaptation Planning Process

1. What is adaptation planning in your perception?
2. Where do you learn it? Do you reference organization or best practices?
3. In which level(s) do you think it should be conducted? Please explain.
4. What kinds of methods that you preferably choose to conduct?
5. How many times do you conduct the planning process, which related to climate change/disaster?
6. What do you think about the validity of data and climate model?
7. Did you feel satisfy with the process? Why?

People's Participation

1. Do you think people need to participate? In what level? Please explain.
2. What do you think about the level participation of people in Jakarta? Penjaringan?
3. What are the useful things that you can get from the people? Knowledge? Experiences?
4. According to you, is it a must that planning should consider people's knowledge? People's experiences? Can you explain it?

Institutions

1. Does Jakarta have regulation or international best practices to be referenced? Does it work?
2. Do you think we need a new rule or regulation to conduct adaptation planning?
3. In the community level, do you think it is necessary to build an organization to plan and implement the adaptation plan?
4. How do you see the community plan to the government's planning, programming, and budgeting?
5. Who do you think are the most important actors in adaptation planning?
6. What kind of resources that can be used at the community level?

Thank you.

Head of RTs/RWs

This guideline will provide semi-structured questions for the selected head of RTs/RWs. This research will investigate mainly the process of adaptation planning, peoples' participation, and the institutions. The questions in here however will be aimed at asking the Head RTs/RWs' perception on vulnerability in their community.

Exposure

1. Could you mention what are the disasters which ever affected your community?
2. Do you think your place is exposed to impact of climate change? Or other hazards?
3. How bad is the impact? Please explain.
4. Do you know which one of your communities whom hit most?
5. What have been done to reduce the exposure? Do you see any difference before and after the development?

Sensitivity

1. Do you agree with some opinions that say your area is poor?
2. Why do you think they are poor?
3. In which area is the poorest household?
4. Do you think there are any factors that make your community sensitive to hazards?

Adaptive Capacity

1. Who are the most suffered persons in your community?
2. Why do you think they suffered?
3. Can you explain who are the persons that can be counted on to respond the disaster, based on previous experiences?
4. What did they do?
5. Do you have any adaptation measures that have been built by your (community) own?

Thank you

Appendix 4: Transcript Form

Interviewer:	Date and Location of Interview
Information of Respondent: Name: Position: Organization:	Contacts of respondent: Email: Phone: Handphone:
Was the interview being recorded? Yes ☐ No ☐	Informed consent was given by respondent: Verbally ☐ Written ☐

Appendix 5: The Occurrence of Tidal Floods (Banjir Rob) from 2007 to 2014

Year	Month/date	Location	Source of information
2007	23-Aug	Muara Baru	www.liputan6.com
	25-Nov	Muara Baru	www.okezone.com
	20-Dec	Muara Baru	www.okezone.com
2008	08-May	Airport area	www.okezone.com
	06-Feb	Penjaringan, Kamal Muara	www.rapi-nusantara.net
	14-Nov	Muara Baru	www.okezone.com
	01-Dec	Muara Baru, Tanjung Priok	www.okezone.com

Year	Month/date	Location	Source of information
	14-Dec	Muara Baru, Tanjung Priok	www.okezone.com
2009	11-Jan	Muara Baru, Penjaringan, Muara Kapuk, Pluit	http://news.viva.co.id
	09-Feb	Ancol, Marunda	http://video.tvonews.tv
	12-May	Kamal Muara, Kapuk, Kapuk Raya	www.okezone.com
	14-Oct	Marunda	http://metro.news.viva.co.id
	19-Oct	Muara Baru	http://metro.news.viva.co.id
	05-Nov	Marunda, Kamal Raya	http://desasejahtera.org
	02-Dec	Tanjung Priok	www.meandmycoastlife.blogspot.com
2010	01-Jan	Tanjung Priok	http://metro.news.viva.co.id
	29-Jan	Pademangan	http://berita.liputan6.com
	13-Feb	Muara Baru	www.beritajakarta.com
	15-Jun	Tanjung Priok	www.meandmycoastlife.blogspot.com
	24-Jun	Muara Baru, Penjaringan	http://m.poskota.co.id
2011	03-Jan	Tanjung Priok	http://megapolitan.kompas.com
	17-Jan	Muara Baru	http://megapolitan.kompas.com
	31-Oct	Tanjung Priok, Muara Baru, Muara Angke	www.mediaindonesia.com
	25-Nov	Pantai Mutiara, Pluit, Penjaringan	http://megapolitan.kompas.com
	23-Dec	Tanjung Priok, Muara Baru	www.detik77.com
2012	19-Jan	Cengkareng, Muara Baru	http://megapolitan.kompas.com
	20-Jan	Tanjung Priok	www.okezone.com
	09-Dec	Kamal Muara	www.status-dunia.blogspot.com
	16-Nov	Kalideres	www.indosiar.com
2013	14-Jun	Sunter Agung, Pademangan	www.republika.co.id
	19-Jun	Ancol	http://megapolitan.kompas.com
	20-Jan	Pluit	www.didisadili.blogspot.com
2014	17-Jan	Pademangan, Ancol, Muara Baru, Penjaringan	http://megapolitan.kompas.com
	29-Jan	Muara Baru, Kamal Muara	http://megapolitan.kompas.com
	02-May	Cilincing	www.fajar18feb.blogsport.com

Source: Author, based on Hiladaliyani (2011)

Appendix 6: Social and Public Facilities in Kelurahan Penjaringan

No.	Facilities	Units	Remarks
1	Mosque and *mushola*	75	–
2	Church	9	–
3	Buddhist praying place	4	–
4	Community health center	2	–
5	Health clinic	11	State and private
6	Elementary school	23	State and private
7	Junior high school	8	Including religious school
8	Senior high school	7	Including religious school
9	University	1	Private
10	International school	1	Private
11	Art and culture place	5	–
12	Tourism and historical object	3	–
13	Sport arena	6	Private
14	Market/shops	3	Private
15	Gas station	5	Pertamina and Shell
16	Police station	4	–
17	Fire station	2	–
18	Flood monitoring post	1	Water pump area
16	Disaster coordinating post	2	Government office building
17	Disposal site	3	–
18	Chinese cemetery	3	–
20	Military district office	1	–
21	Neighborhood security post	240	–

Appendix 7: Superimpose Analysis with ArcGIS 9.0

Spatial analysis is a combining process of several information based on various layers of data by using certain spatial operation. It is a process that answers the question related to the spatial information. We can use vector, satellite imagery, and tabular data. There are at least five steps to do the spatial analysis: defining key questions, collecting and preparing data, choosing the method or analytical tools, analytical process, and checking and editing the results.

Picture 1 ArcGIS toolbox

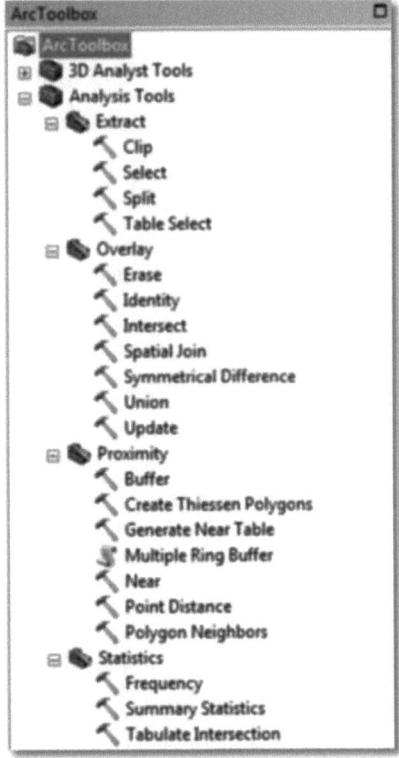

Picture 2 Overlay process

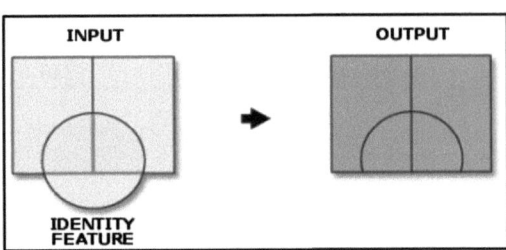

Analytical Tools Using ArcGIS

ArcGIS has several techniques in the arc toolbox (see Pictures 1 and 2) to analyze spatial data, which consists of:

- *Extraction*, which consists of clip, select, split, and table select.
- *Overlay*, which consists of erase, identity, intersection, symmetrical difference, union, and update.
- *Proximity*, which consists of buffer, multiple ring buffers, and near and point distance.

- *Statistic*, which consists of frequency and summary statistic. In this research, I use the intersect tool because this method superimposes of the vector data of the identity feature to the input area and allows the integration of tabular data in the input features.

Data Processing

Before I preceded the overlay process, I had collected various maps that can represent the flood and poor in *Kelurahan Penjaringan*. Based on my secondary data collection, I have listed three kinds of flood map that represent flood as following:

1. The distribution of flooded area based on the big flood occurrences in year 2002 and 2007, which was released by the Government DKI Jakarta province. It would show the facts on which area of Penjaringan has been inundated during the 2002 and 2007 floods (see Picture 4).
2. The map of flood scenario by 2030 that was predicted by the government of DKI Jakarta province. It would represent the future prediction of floods (see Picture 5). The scenario was built based on the 2007 flood and was divided into four classes: 0–10 cm, 10–30 cm, 30–100 cm, and 1–5 m (see Picture 6).
3. The flood hazard map based on the historical data of flood events during the last 20 years, which was recorded by BNPB. It would represent the potential risks of flood that classified into two classes, high and very high (see Picture 7), since all Kelurahan Penjaringan has been categorized as the most risk area.
4. The map of recurrence interval of flood inundation that represents the probability of floods in Kelurahan Penjaringan. The map was adopted from the World Bank Indonesia that categorized the interval into three levels, first is 1–2 years, second is 5–10 years, and third is 50–100 years interval (see Picture 8),

and one map, the slum distribution that represents the distribution of poor households in *Kelurahan Penjaringan*. The tabular data was divided into four levels, heavy, medium, slightly, and not slum, using the slum categorization of Government of DKI Jakarta Province (see Picture 8).

In order to operate this tool, firstly, I selected the identity options in the arc toolbox of the ArcGIS 9.1 software. I inputted the five maps above as the features that need to be overlaid. I only selected the polygon area as joined attributes (see Picture 3) and used the RW boundary as the unit analysis. This overlay process could show which RW has the most superimposed data. The result identified one of the RWs that can represent the area where interplay between floods and the poor occurred.

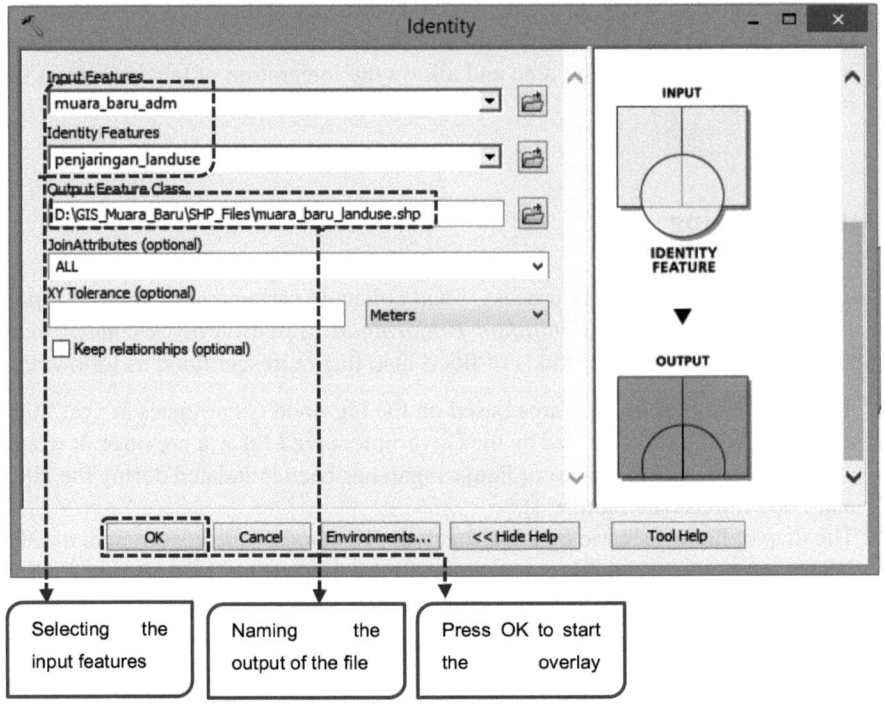

Picture 3 Overlay process

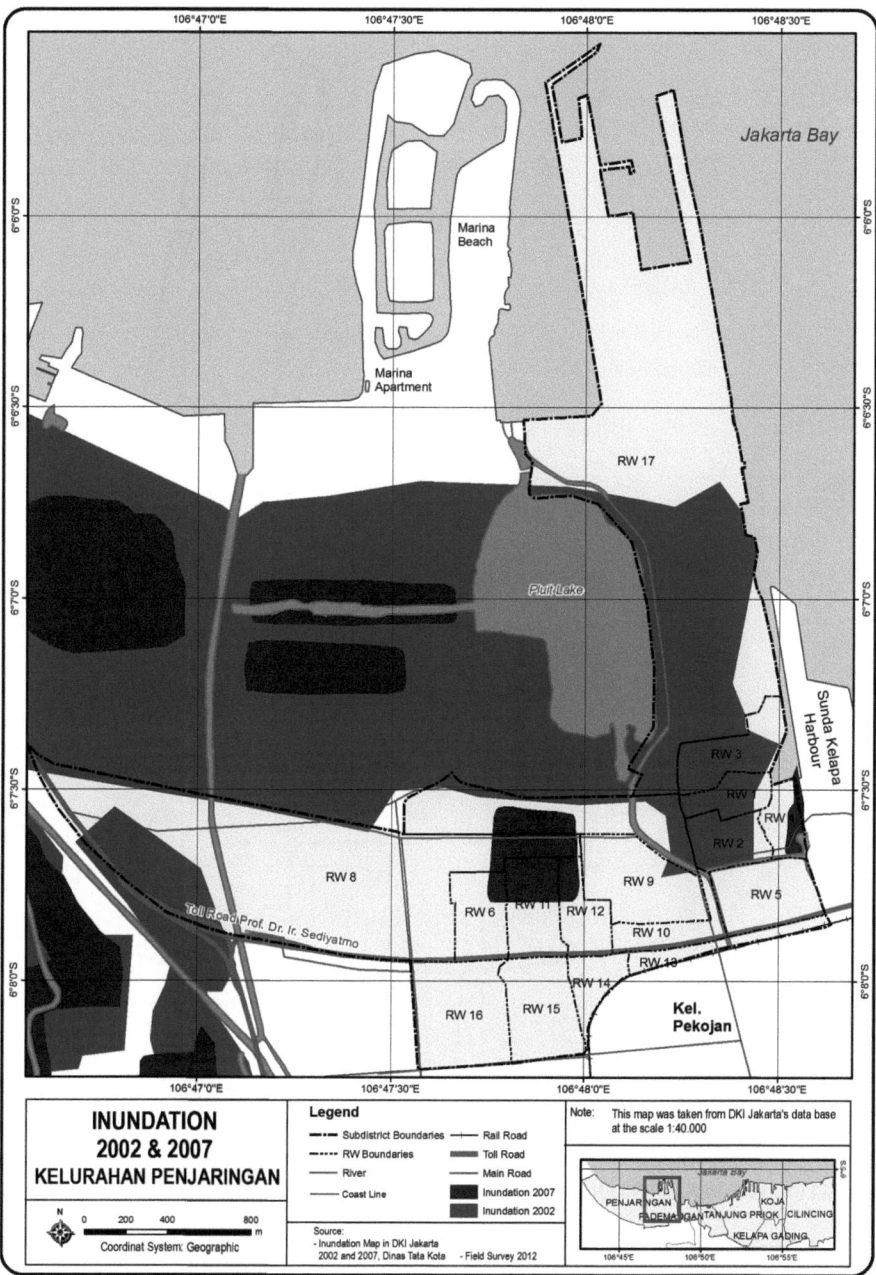

Picture 4 Map of flood Inundation of 2002 and 2007 (Source: Government of DKI Jakarta, 2009)

Picture 5 Flood scenario by 2030 (Government of DKI Jakarta, 2009)

Appendices 181

Picture 6 Flood hazard (Source: BNPB, 2009)

Picture 7 The recurrence interval of flood inundation (Source: World Bank Indonesia, 2007)

Appendices 183

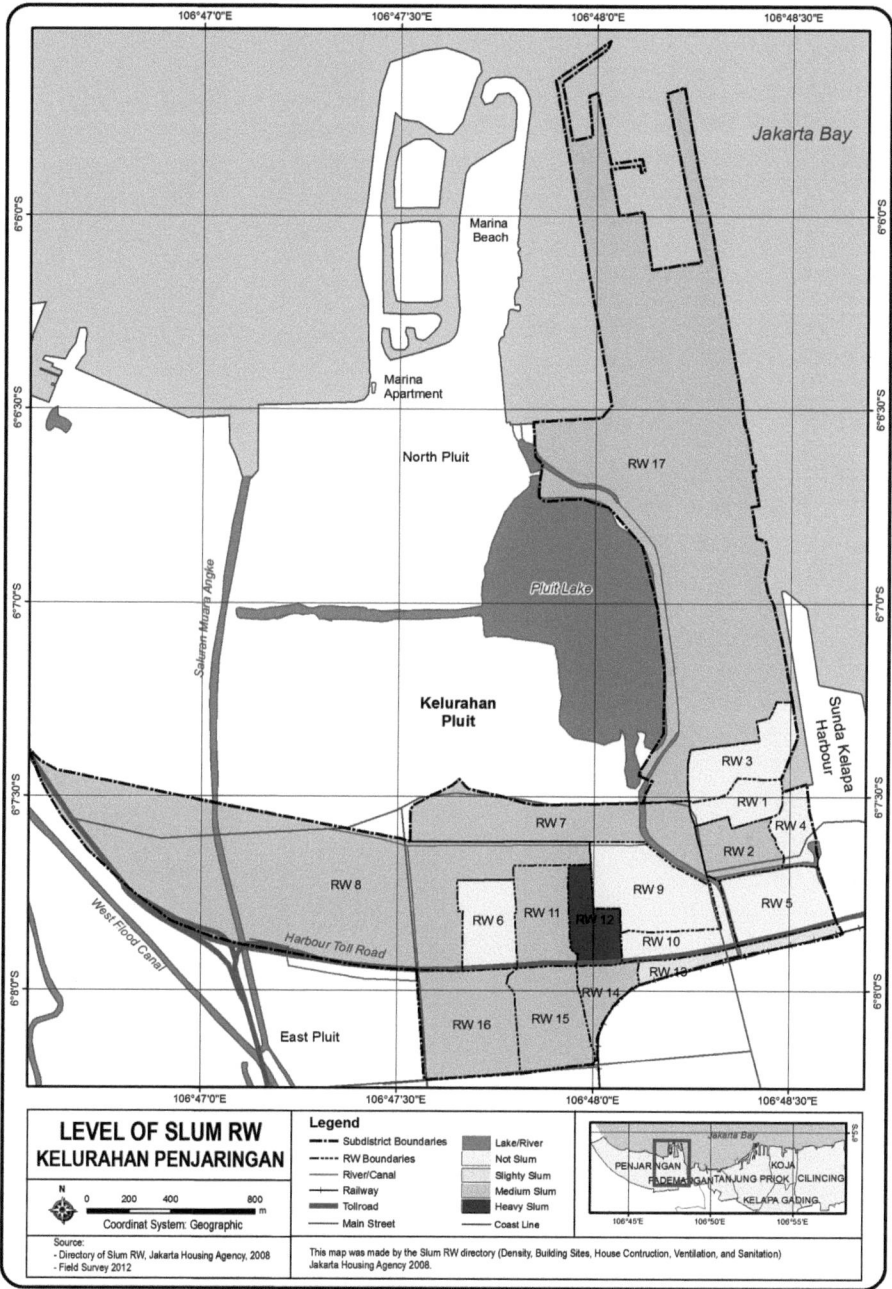

Picture 8 Distribution of the poor in Penjaringan

Glossary

Aqua	Water bottle
Arisan	A form of Rotating Savings and Credit Association in Indonesian culture, a form of Microfinance
Bajaj	Three-wheeled vehicle, used for transporting a maximum of only four people
Bakso	Meatballs
Banjir	Flood
Banjir biasa	Rainy flood
Banjir kiriman	Transferred flood
Banjir Rob	Tidal flood
Bedeng	Temporary house or shelter
Buah	Fruits
Curhat	Private talk, telling story deeply
Desa Tangguh	Resilient village
Dinas Tata Ruang	Urban Planning Department
Idul Fitri	Eid Mubarak—Traditional Muslim greeting reserved for use on the festivals of Eid al-Adha and Eid al-Fitr
Gak apa-apa	No problem
Gang	Alley, small road
Genset	Generator
Girik	Evidence of land controlling and tax payment on parcel of land and building on it (if any)
Gorengan	Fried food
Gotong royong	Mutual aid or cooperation in the community
Hijriah Calendar	The lunar year numbering system used in the Islamic calendar
Ibu	Mrs.
Ibu-ibu	Women

Indomie	Instant noodle
Isra Miraj	Muslim day
Jamu	Traditional herb products
Jaya	Glory
Kali	River
Kampung	Village
Kampung deret	Kampung readjusment program
Kanem (java)	The sixth
Kapitu (java)	The seventh
Katulampa	A Watergate's name in Bogor
Kawolu (java)	The eighth
Kecamatan	Subdistrict
Kelurahan	Village, a part of regency/city administrative area
Kentongan	Traditional tool made by wood for indicating the clock or emergency
Kerja bakti	Voluntary communal work
Ketua RT	Head of RT
Ketua RW	Head of RW
Kotak	Box
Laut	Sea
Lurah	Head of Kelurahan
Maghrib	Praying time for Muslim, usually between 6:00 pm and 6:30 pm
Majelis Taklim	Muslim solidarity
Manusia perahu	Boatman
Musibah	Disaster
Musrenbang	Planning forum
Orde Baru	The New Order (1966–1998)
Orde Lama	The Old Order (1945–1965)
Pengajian	Al-Quran reading
Peraturan Gubernur	Governor Regulation
Perda	Regional Regulation
Perwakilan	Representative
Pronas	National program in the economic crisis era that prohibited people to occupy abandoned land
Rembuk	Meeting, people gathered talking about an issue
Revitalisasi Pantura Jakarta	The planning study for new development of Jakarta coastal area
Rukun Tetangga	Neighborhood
Rukun Warga	Community Association
Rusunami	Cheap apartment for low-income dwellers
Sasi (Papuan)	Papuan community who use their traditional tenure management system
Sholat	Praying
Sudah dari sononya	Naturally, given

Tanggul	Sea dyke
Tenggelam	Drowned
Teras	Terrace, front floor
Universitas	University
Waduk	Reservoir
Warung	Small shop
Warung nasi	Small shop selling food, mainly rice
ber and *ri*	Indonesian suffixes for months of rainy season

References

Aassche, K. V. (2007). Planning as/and/in context: Towards a new analysis of context in interactive planning. *Journal of METU JFA, 2*(24), 2.

Abidin, H. Z., Andreas, H., Djaja, R., Dudy, D., & Gamal, M. (2008). Land subsidence characteristics of Jakarta between 1997 and 2005, as estimated using GPS surveys. *GPS Solutions, 12*(1), 23–32.

ACF – Action Contre la Faim. (2010). *Manajemen Bencana di Tingkat Lokal: Formalisasi SATLINMAS & SPTB*. Jakarta: ACF.

ADB – Asian Development Bank. (2009). *Adaptation assessment, planning and practice: An overview from the Nairobi work programme on impacts, vulnerability and adaptation to climate change*. In United Nations Framework Convention on Climate Change (Ed.). Bonn: UNFCCC.

ADB – Asian Development Bank. (2011). *Community – based climate vulnerability assessment and adaptation planning: A Cook Islands Pilot Project*. Manila: Asian Development Bank.

Adger, W. N., & Kelly, P. M. (1999). Social vulnerability to climate change and the architecture of entitlements. *Mitigation and Adaptation Strategies for Global Change, 4*(3–4), 253–266.

Adger, W. N., Huq, S., Brown, K., Conway, D., & Hulme, M. (2003). Adaptation to climate change in developing world. *Progress in Development Studies, 3*(3), 179–195.

Adger, W. N., Dessai, S., Goulden, M., Hulme, M., Lorenzoni, I., Nelson, D. R., et al. (2009). Are there social limits to adaptation to climate change? *Climate Change, 93*, 335–354.

AICP – American Institute of Certified Planners. (2013). *History and theory of planning: Why we do and what we do* [Review of Georgia Planning Association AICP Exam, January 18]. http://georgiaplanning.org/presentations/AICP_exam_reviews/2013_AICP%20lecture%20on%20theory%20and%20history.pdf

AKP – Adaptation Knowledge Platform. (2013). *Understanding adaptation planning: Selected case studies in Nepal, Philippines and Vietnam, adaptation knowledge platform*. Partner Report Series No. 9. Bangkok: Stockholm Environment Institute. http://www.asiapacificadapt.net or weADAPT.org

Akuntono, I. (2013, January 21). *Bantuan untuk korban banjir di Muara Baru masih minim*. Kompas.

Alam, M., & Murray, L. A. (2005). Facing up to climate change in South Asia. In *IIED gatekeeper series 118*. London: IIED.

Aldrian, E. (2007). *Panduan Simulasi Sea-Air Interaction dengan Couple Regional Climate Model*. Unit Pelaksana Teknis Hujan Buatan, Badan Pengkajian dan Penerapan Teknologi. Jakarta: BPPT.

Alexander, E. A. (1992). *Approaches to planning* (2nd ed.). Philadelphia: Gordon and Breach Science Publishers.

Alexander, E. A. (2005). Institutional transformation and planning: From institutionalization theory to institutional design. *Journal of Planning Theory, 4*(3), 209–223.

AlSayyad, N., & Roy, A. (2004). *Urban informality: Transnational perspectives from the Middle East, Latin America, and South Asia*. Washington, DC: Lexington Books.

Alwang, J., Siegel, P. B., & Jorgensen, S. L. (2001). *Vulnerability: A view from different disciplines, Social protection discussion paper series*. Washington DC: The World Bank.

AMS – American Meteorological Society. (2012). Glossary of meteorology. http://glossary.ametsoc.org/wiki/Flood. Accessed 23 Mar 2012.

Antweiler, C. (2004). Local knowledge theory and methods: An urban model from Indonesia. In A. Bicker, P. Sillitoe, & J. Pottier (Eds.), *Investigating local knowledge: New directions, new approaches*. Aldershot: Ashgate.

Antweiler, C. (2012). Urban environmental knowledge as citizen's science: Theoretical issues and methods from indonesia. In A.-k. Hornidge & C. Antweiler (Eds.), *Environmental uncertainty and local knowledge: Southeast Asia as a laboratory of global ecological change*. Majuske Medienproduktion GmbH: Bielefeld.

Antweiler, C., & Mersmann, C. (1996). Local knowledge and cultural skills as resources for sustainable forest development. Bundesministerium für wirtschaftliche Zusammenarbeit und Entwicklung (BMZ) and Deutsche Gesellschaft für Technische Zusammenarbeit (GTZ), Eschborn.

Argyris, C., & Schon, D. A. (1978). *Organizational learning: A theory of action approach*. Reading: Addision Wesley.

Aziza, K.S. (2014, January 14). *Jakarta Darurat Banjir*, BNPB dan BPPT tebar garam. http://tekno.kompas.com/read/2014/01/14/1449508/jakarta.darurat.banjir.bnpb.dan.bppt.tebar.garam. Accessed 14 Jan 2014.

Babbie, E. (1995). *The practice of social research (7th edition)*. Belmont, CA: Wadsworth.

Bahrainy, H. (1998). Urban planning and design in a seismic-prone region (Rasht in Northern Iran). *Journal of Urban Planning and Development, 124*(4), 148–181. doi: http://dx.doi.org/10.1061/(ASCE)0733-9488(1998)124:4(148)

Bankoff, G. (2004). *The historical geography of disaster: "vulnerability" and "local knowledge" in Western Discourse*. London: Earthscan.

Banks, N., Roy, M., & Hulme, D. (2011). Neglecting the urban poor in Bangladesh: Research policy and action in the context of climate change. *Environment and Urbanization, 23*, 487. doi:10.1177/0956247811417794.

Barber, M. (2014). Alfred Schütz, The Stanford Encyclopedia (Spring Edition). In Edward N. Salta (Ed.)

Baum, H. S. (1980). Sensitizing planners to organization. In P. Clavel, J. Forester, & W. Goldsmith (Eds.), *Urban and regional planning in an age of austerity*. New York: Pergamon Press.

Bayat, A. (2000). *Social movements, activism and social development in the Middle East*. Geneva: UNRISD.

BBC – British Broadcast Company. (2011, November 9). Vietnam floods crisis deepens, with deaths topping 100. BBC. http://www.bbc.com/news/world-asia-15650162. Accessed December 2011.

Beard, V. A. (2002). Covert planning for social transformation in Indonesia. *Journal of Planning Education and Research, 22*(1), 15–25.

Belarminus, R. (2013, Januari 23). Tanpa Pompa, banjir di pluit baru surut 7 hari. *Kompas Online*. http://megapolitan.kompas.com/read/2013/01/20/22123562/Tanpa.Pompa.Banjir.di.Pluit.Baru.Surut.7.Hari. Accessed Feb 2013.

Bengtsson, J., Hargreaves, R., & Page, I. C. (2007). *Assessment of the need to adapt buildings in New Zealand to the impacts of climate change* (p. 179). Porirua: Branz Study Report SR.

Berger, P., & Luckmann, T. (1967). *The social construction of reality: A treatise in the sociology of knowledge*. Harmondsworth: Penguin Group.

Bernard, R.H. (2006). *Research Methods in Anthropology: Qualitative and Quantitative Approaches* (Vol. 4th Ed.). UK: Alta Mira Press.

Berner, E. (2000). *Informal developers, patrons, and the state: Institutions and regulatory mechanism in popular housing*. Paper presented at the ESF/N-AERUS Coping with informality and illegality in human settlements in developing cities Workshop, Leuven.

Birkmann, J. (2005). Danger need not spell disaster – But how vulnerable are we? In *Research brief*. Tokyo: United Nations University.

Birkmann, J., & Fernando, N. (2008). Measuring revealed and emergent vulnerabilities of coastal communities to tsunami in Sri Lanka. *Disasters, 32*(1), 82–105.

Birkmann, J., Garschagen, M. A., Krass, F., & Quang, N. (2010). Adaptive urban governance-new challenges for the second generation of urban adaptation strategies to climate change. *Sustainability Science, 5*(2), 185–206.

Blakely, E. J. (2007). *Urban planning for climate change*. Cambridge: Lincoln Institute of Land Policy Working Paper.

BNPB – Badan Nasional Penanggulangan Bencana. (2012). *Rencana Penanggulangan Bencana Provinsi DKI Jakarta: Laporan Akhir*, Desember 2012. Jakarta.

BNPB – Badan Nasional Penanggulangan Bencana. (2014). *Data Dijital Bencana Indonesia (DIBI)*. Jakarta: BNPB.

Boer, Rizaldi. (2013). *Penyusunan Rencana Aksi Daerah Adaptasi Perubahan Iklim: Laporan Kemajuan*. DKI Jakarta.

Bogardi, J. J., Villagran, J., Birkmann, J., Renaud, F., Sakulski, D., Chen, X., et al. (2005). Vulnerability in the context of climate change. *Human security and climate change, an international workshop Holmen fjord hotel*, Asker.

Bohle, H. G. (2001). Vulnerability and criticality: Perspectives from social geography. *IHDP Update, 2*(2001), 1–7.

Bollin, C., Cárdenas, C., Hahn, H., & Vatsa, K. S. (2003). *Disaster risk management by communities and local governments*. Inter-American Development Bank.

Bosher, L., Penningrowsell, E., & Tapsell, S. (2007). Resource accessibility and vulnerability in Andhra Pradesh: Caste and non-caste influences. *Development and Change, 38*, 615–640.

Bowo, F. (2010). *Toward sustainable Jakarta in the time of climate change*. Paper Presented at World Urban Forum (June 23), Singapore.

BPS-Biro Pusat Statistik. (2010). *Preliminary results of population census*. Jakarta: Bureau of Statistics, Ministry of Finance and Development Planning.

BPS – Biro Pusat Statistik. (2012). *Jakarta Dalam Angka, Jakarta in figures*. Jakarta: BPS Provinsi DKI Jakarta.

BPS – Biro Pusat Statistik DKI Jakarta. (2013). *Jakarta Dalam Angka, Jakarta in figures*. Jakarta: BPS Provinsi DKI Jakarta.

Briassoulis, H. (1997). How the others plan: Exploring the shape and forms of informal planning. *Journal of Planning Education and Research, 17*(2), 105–117.

Brinkman, J.J. (2012, June). *Jakarta coastal defense strategy (JCDS): Sinking cities-Jakarta*. Paper Presented on APA Conference, Los Angeles.

BRKP – Badan Riset Kelautan dan Perikanan-RI. (2009). *The arrangement of marine information for Copenhagen meeting*. Executive summary. Jakarta.

Brown, O. (2008). *Migration and climate change*. Geneva: International Organization for Migration.

Bunnell, T., Ho, K. C., & Drummond, L. (2002). *Critical reflections on cities in Southeast Asia*. Leiden/Singapore: Times Academic Press/ Brill Academic Publishers.

Burton, I., Huq, S., Lim, B., Pilifosofa, O., & Skipper, L. (2002). From impacts assessment to adaptation priorities: The shaping of adaptation policy. *Climate Policy, 17*, 145–159.

Caeli, K. (2001). Engaging with phenomenology: Is it more of a challenge than it needs to be?. *Quantitative Health Research*, 11, 273–282.

Canadian Institute of Planners (CIP). (2011). *Climate change adaptation planning: A handbook for small Canadian communities*. CIP. Available at https://www.cip-icu.ca/Files/Resources/RURAL-HANDBOOK-FINALCOPY

Cannon, T., Twigg, J., Rowell, J. (2003). Social vulnerability. Sustainable livelihoods and disasters. Report to DFID Conflict and Humanitarian Assistance Department (CHAD) and

Sustainable Livelihoods Support Office. http://www.benfieldhrc.org/disaster_studies/projects/soc_vuln_sust_live.pdf.
Cardona, O. D. (2004). The need for rethinking the concepts of vulnerability and risk from a holistic perspective: A necessary review and criticism for effective risk management. In *Mapping vulnerability: Disasters, development and people* (pp. 37–51).
CARE. (2010). *Community-based adaptation toolkit*. Digital toolkit (Version 1.0) July 2010.
CARE. (2013). *Community based resilience assessment (CoBRA) conceptual framework and methodology*. Dry lands Development Centre. UNDP.
Carmin, J.A. (2009). *Planning climate resilience cities: Early lessons from early adapters. Marseille, France*. Paper presented for the World Bank, 5th Urban Research Symposium, Cities, and Climate Change.
Carmin, J., Anguelovski, I., & Roberts, D. (2012). Urban climate adaptation in the global South planning in an emerging policy domain. *Journal of Planning Education and Research, 32*(1), 18–32.
Carter, T. R. (1996). Assessing climate change adaptations: the IPCC guidelines. In J. Smith, N. Bhatti, G. Menzhulin, R. Benioff, M. I. Budyko, M. Campos, B. Jallow, & F. Rijsberman (Eds.), *Adapting to climate change: An international perspective* (pp. 27–43). New York: Springer.
CCCD, Commission on Climate Change and Development. (2009). *The human dimension of climate adaptation: the importance of local and institutional issues*. In Ian Christoplos, Simon Anderson, Margaret, Victor Galaz, Merylyn Hedger, Richard J.T. Klein, & Katell Le Goulven (Ed.).
Chambers, R., & Conway, G. (1992). *Sustainable rural livelihoods: Practical concepts for the 21st century*: Institute of Development Studies (UK).
Choi, O., & Fisher, A. (2003). The impacts of socioeconomic development and climate change on severe weather catastrophe losses: Mid-Atlantic region (MAR) and the US. *Climatic Change, 58*(1–2), 149–170.
Christmann, G., Ibert, O., Kilper, H., & Moss, T.U. (2012). Vulnerability and Resilience from a Socio-Spatial Perspective, Towards a Theoretical Framework. *Working Paper, Erkner, Leibniz Institute for Regional Development and Structural Planning.*
Christoplos, I., Anderson, S., Arnold, M., Galaz, V., Hedger, M., Klein, R. J., & Goulven, K. (2009). *The human dimension of climate adaptation: The importance of local and institutional issues*. Stockholm: Commission on Climate Change and Development.
CIP- Canadian Institute of Planner. (2011). *Climate change adaptation planning: A handbook for small Canadian communities*. Ottawa: CIP.
Colbran, N. (2009). Will Jakarta be the next Atlantis? Excessive groundwater use resulting from a failing piped water network. *Environment and Development Journal, 18*.
Conway, M. (2003, September 28). *An introduction to scenario planning*. Paper Presented for a Foresight Methodologies Workshop, Thinking Futures, Australia.
Conway, M. (2007, November). *Building a strategic foresight capacity*. Paper Presented at AAIR Forum, Thinking Future, Australia.
Creswell, J.W. (1994). *Qualitative and quantitative approaches*. Thousand Oaks: Sage.
Creswell, J.W. (1998). Qualitative inquiry and research design: Choosing among five traditions. Sage.
CUS – Centre for Urban Studies. (2006). *Slums of urban Bangladesh Dhaka*. Dhaka: CUS/MEASURE/NIPORT/USAID.
Dalton, L. C. (1989). Emerging knowledge about planning practice. *Journal of Planning Education and Research, 9*(1), 29–44.
Damar, A. (2003). *Effects of enrichment on nutrient dynamics, phytoplankton dynamics and productivity in Indonesian tropical waters: A comparison between Jakarta Bay, Lampung Bay and Semangka Bay*. PhD dissertation, Christian-Albrechts-Universität.
Damm, M. (2010). *Mapping Social-Ecological Vulnerability to Flooding*. Graduate Research Series, PhD Dissertations Publication Series of UNU-EHS, Vol. 3.
Davidson, R. (1997). An urban earthquake disaster risk index. John A. Blume Earthquake Engineering Center Technical Report Series. Stanford.

Denhardt, R. B. (1981). *The shadow of organization*. Lawrence: The Regents Press of Kansas.

Dewalt, K.M., & Dewalt B.R. (2002). *Participant observation: A guide for fieldworkers*. Walnut Creek: Altamira Press.

Dewey, J. (1998). *Experience and Education: The 60th anniversary edition, kappa delta pi*. West Lafayette: International Honor Society in Education.

DFID – The Department for International Development of UK. (2005). *Natural disaster and disaster risk reduction measures: A desk review of costs and benefits*. London: UK Department for International Development, The Stationery Office.

DJPR. (2013). Pembangunan dan Pengembangan Kota Pesisir berbasis 'Kota Hijau'. Bahan Presentasi pada acara Kongres ASPI VII dan Seminar Nasional Kota Hijau Pesisir Tropis, Universitas Sam Ratulangi. Manado.

DKI Jakarta – Government of DKI Jakarta Province. (2011). *Rencana Tata Ruang Wilayah DKI Jakarta 2010–2030*. Draft Perda. DKI Jakarta: Bappeda.

Droesch, A.C., Gaseb, N., Kurukulasuriya, P., Mershon, A., Moussa, K.M., Rankine, D., & Santos, A. (2008). A guide to the vulnerability reduction assessment. Community-Based Adaptation Programme, UNDP Working Paper.

Durand-Lasserve, A. (1998). Law and urban change in developing countries: Trends and issues. In E. Fernandes & A. Verley (Eds.), *Ilegal cities - law and urban change in developing countries* (pp. 233–257). London: Zed Books.

Dwianto, R. D. (2003). An existing form of urban locality groups in Jakarta: Reexamining RT/RW for the post- New Order Era. In *Representing local places and raising voices from below – Japanese contributions to the history of geographical thought*.

Ehlert, J. (2012). We observe the weather because we are farmers: Weather knowledge and meteorology in the Mekong Delta, Vietnam. In A.-K. Hornidge & C. Antweiler (Eds.), *Environmental uncertainty and local knowledge Southeast Asia as a Laboratory of Global Ecological Change*. Biefeld: Verlag.

Eliasson, I. (2000). The use of climate knowledge in urban planning. *Landscape and Urban Planning, 48*(1), 31–44.

Elings, C.F. (2011). Jakarta climate adaptation tools: Tools and capacity building for climate proofing of integrated spatial planning and flood risk management. *Research of Delta Alliance*. Netherlands: Royal Haskoning SMC.

Fainstein, S. S. (2005). Planning theory and the city. *Journal of Planning Education and Research, 25*(2), 121–130.

Fainstein, S. S., & Campbell, S. (2003). *Readings in planning theory*. Oxford: Wiley-Blackwell.

Fay, B. (1987). *Critical social science*. Cambridge: Polity Press.

Fekete, A. (2010). *Assessment of social vulnerability to river floods in Germany*. In Graduate Research Series, PhD Dissertations: Publication Series of UNU-EHS.

Fernandes, E. (1994). Defending collective interests in Brazilian environmental law: An assessment of the civil public action. *Review of European Community and International Environmental Law, 3*(4), 253–258.

Firman, T., Suroso, D., Sofian, I. (2009). Sea level rise in Java's north coast region: Implication for urban development policy and planning. Paper presented at cities at risk conference-2, Oktober 2009. Bangkok.

Firman, T., Surbakti, I. M., Idroes, I. C., & Simarmata, H. A. (2011). Potential climate-change related vulnerabilities in Jakarta: Challenges and current status. *Habitat International, 35*(2), 372–378.

Fischler, R. (2012). Reflective practice. In B. Sanyal, L. Vale, & C. Rosan (Eds.), *Planning ideas that matter*. Cambridge, MA: MIT Press.

Fitrinitia, I. S. (2012). Adaptation startegy of coastal communities related to disaster management, case study: Muara Baru and Pluit community to flood in DKI Jakarta province. Master Thesis, Sociology Department, Universitas Indonesia.

Flick, U., Kardorff, E. V., & Steinke, I. (2004). *A companion to qualitative research*. London: Sage.

Flood, R. L., & Romm, N. R. A. (1996). *Diversity management: Triple loop learning*. Chicester: Wiley.

Folke, C., Chapin, F. S., III, & Olsson, P. (2009). Transformations in ecosystem stewardship. In G. P. Kofinas, C. Folke, & F. S. Chapin III (Eds.), *Principles of ecosystem stewardship: Resilience-based natural resource management in a changing world*. New York: Springer.

Forester, J., & Stitzel, D. (1989). Beyond neutrality. *Negotiation Journal, 5*(3), 251–264.

Forester, J., Clavel, P., & Goldsmith, W. (1980). *Urban and regional planning in an age of austerity*.

Friedmann, J. (1969). Planning and societal action. *Journal of the American Institute of Planners, 35*, 311–318.

Friedmann, J. (1984). Agropolitan development: A territorial approach to meeting basic needs. In D. Korten & R. Klauss (Eds.), *People center development: Contributions towards theory and planning framework* (pp. 210–239). West Hartford: Kumarian Press.

Friedmann, J. (1987). *Planning in the public domain: From knowledge to action*. Princeton: Princeton University Press.

Friedmann, J. (2000). The good city: In defense of utopian thinking. *International Journal of Urban and Regional Research, 24*(2), 460–472.

Friedmann, J. (2008). The uses of planning theory a bibliographic essay. *Journal of Planning Education and Research, 28*(2), 247–257.

Fuchs, R. J. (2010. *Cities at risks: Asia's coastal cities in an age of climate change*. Working Paper July 2010. Honolulu: East-West Center.

Füssel, H.-M. (2007). Adaptation planning for climate change: Concepts, assessment approaches, and key lessons. *Sustainability Science, 2*(2), 265–275.

Füssel, H.-M., & Klein, R. J. T. (2006). Climate change vulnerability assessments: An evolution of conceptual thinking. *Climate Change Journal, 72*(3), 301–329.

Gaillard, J.-C. (2010). Vulnerability, capacity and resilience: Perspectives for climate and development policy. *Journal of International Development, 22*(2), 218–232.

Garison Institute (2013). *The human dimensions of resilient and sustainable cities: Exploring strategies for working with people, organizations, and social networks to forestall crises and enhance the ability of cities to bounce back from crises*. Climate, Cities, and Behaviour Symposium: Synthesis Report. Garrison Institute.

Garr, D. (1989). Indonesia's Kampung improvement program: Policy issues and local impacts for secondary cities. *Journal of Planning Education and Research, 9*, 79–83.

Garzón, C., Cooley, H., Heberger, M., Moore, E., Allen, L., Matalon, E., Doty, A., & The Oakland Climate Action Coalition. (2012). *Community- based climate adaptation planning: Case study of Oakland, California*. California Energy Commission. Publication number: CEC-500-2012-038.

Gehl, J. (2011). *Cities for people*. Washington, DC: Island press.

Geis, D. E. (2000). By design: The disaster resistant and quality-of-life community. *Natural Hazards Review, 1*(3), 151–160.

Giddens, A. (1984). *The constitution of society: Outline of the theory of structuration*. Oakland: University of California Press.

Goffman, E. (1959). *The presentation of self in everyday life*. New York: Anchor Books.

Gran, G. (1983). *Development by people citizen construction of a just world*. New York: Praeger Publishers.

Granberg, M., & Elander, I. (2007). Local governance and climate change: Reflections on the Swedish experience. *Local Environment, 12*(5), 537–548.

Groenewald, T. (2004). A phenomenological research design illustrated. *International Journal of Qualitative Methods (Artikel 4), 3*(1). Retrieved March 23, 2013, from http://www.ualberta.ca/~iiqm/backissues/3_1/html/groenewald.html

Gumilar, I., Abidin, H. Z., Andreas, H., Mahendra, A. D., Sidiq, T. P., & Gamal, M. (2009). Studi Potensi Kerugian Ekonomi (Economic losses) Akibat Penurunan Muka Tanah (Studi kasus: Wilayah Jakarta, Bandung, dan Semarang). Kelompok Keilmuan Geodesi, Institut Teknologi Bandung.

Haan, L. D., & Zoomers, A. (2005). Exploring the frontier of livelihoods research. *Development and Change, 36*(1), 27–47.

References

Habermas, J. (1984). *Theory of communicative action volume one: Reason and the rationalization of society (book)*. (Thomas A. McCarthy, Trans). Boston, MA: Beacon Press. ISBN 978-0-8070-1507-0.
Habermas, J. (1987). Theory of communicative action volume two: Lifeworld and system. In *A critique of functionalist reason* (Book) (Translated by T. A. McCarthy). Boston: Beacon Press. ISBN 0-8070-1401-X.
Hadi, S. (2010, September 25). Tahun 2050 Jakarta Diprediksi Tenggelam. *Republika Online* 25 September 2010. http://www.republika.co.id/berita/breakingnews/metropolitan/10/09/25/136420-tahun-2050-jakarta-diprediksi-tenggelam . Accessed 24 Mar 2012.
Hallegatte, S., Green, C., Nicholls, R. J., & Corfee-Morlot, J. (2013). Future flood losses in major coastal cities. *Nature Climate Change, 3*(9), 802–806.
Harvey, R. (2011, October 20). *Thailand floods: Bangkok 'impossible to protect'*. BBC-British Broadcast Company. http://www.bbc.com/news/world-asia-pacific-15381227. Accessed Feb 2013.
Haryanto, B. (2009, December 3). Climate change and public health in Indonesia impacts and adaptation. Paper presented at Austral policy forum 09-05S. http://nautilus.org/wp-content/uploads/2012/02/haryanto1.pdf
Hasan, A. (2004). The changing nature of the informal sector in Karachi due to global restructing and liberalisation and its repercussions. In A. Roy & N. Alsayyad (Eds.), *Urban informality*. Lanham: Lexington Books.
Healey, P. (2007). The new institutionalism and the transformative goals of planning. *Institutions and Planning*, 61–87.
Heelan, P. A. (2001). The lifeworld and scientific interpretation. *Handbook of phenomenology and medicine* (pp. 47–66). Springer.
Hewitt, K. (1997). *Regions of risk: A geographical introduction to disasters*. Essex: Longman.
Hildaliyani, U. (2011). *Analisis Daerah Genangan Banjir Rob (Pasang) Di Pesisir Utara Jakarta Menggunakan Data Citra Satelit Spot Dan Alos*. Skripsi, Departemen Geofisika dan Meteorologi Fakultas Matematika dan Ilmu Pengetahuan Alam Institut Pertanian Bogor, IPB.
Hillier, J. (2003). Agon'izing over consensus: Why Habermasian ideals cannot be 'real'. *Planning Theory, 2*(1), 37–59. doi:10.1177/1473095203002001005.
Hodgson, G. (2006). What are institutions?. *Journal of Economic Issues XL No 1 March*.
Hofmann, P., Strobl, J., Blaschke, T., & Kux, H. (2008). Detecting informal settlements from Quickbird data in Rio de Janeiro using an object based approach. In *Object-based image analysis* (pp. 531–553). Springer.
Hornidge, A. K. (2011). *Institutions: Conceptual food for thought*. Paper Presented at the Disciplinary Doctoral Program. ZEF, December 5, Bonn.
Hornidge, A.-K., & Antweiler, C. (2012). In A.-K. Hornidge & C. Antweiler (Eds.), *Environmental uncertainty and local knowledge: Southeast Asia as a laboratory of global ecological change*. Bielefeld: Transcript Bielefeld.
Hornidge, A.-K., & Scholtes, F. (2012). Climate change and everyday life in toineke village, west timor: Uncertainties, knowledge and adaptation. *Sociologus, 61*(2), 151–175.
Howe, J., & Langdon, C. (2002). Towards a reflexive. *Planning Theory, 1*(3), 209–225. doi:10.1177/147309520200100302.
Huber, H., Nolan C. K., Heschel, M. S., von Wettberg, E. J., Banta, J., Leuck, A. M., & Schmitt, J. (2004). Frequency and microenvironmental pattern of selection on plastic shade-avoidance traits in a natural population of Impatiens capensis. *The American Naturalist, 163*(4), 548–563.
Hummel, R. P. (1982). *Phenomenology in planning, policy, & administration*. JSTOR.
Huxley, M. (2000). The limits to communicative planning. *Journal of Planning Education and Research, 19*(4), 369–377.
Hycner, R.H. (1999). Some guidelines for phenomenological analysis of interview data. In A. Bruman & R.G. Burgess (Ed.), *Qualitative research Vol.3*. London: Sage.
IAP – Indonesian Association of Planner. (2012). *Planning for integrated coastal adaptation strategy of north coastal Jakarta*. Final Report. Unpublished, Jakarta.

IIED – International Institute for Environment and Development. (2011). *Climate change, disaster risk, and the urban poor: Cities building Resilience for a changing world.* Washington: World Bank.

IIED – International Institute for Environment and Development. (2012). *Rethinking finance for development the Asian Coalition for Community Action (ACCA).* London: IIED.

IISD – International Institute for Sustainable Development. (2012). *CRiSTAL user manual version 5: Community-based risk screening tool-adaptation and livelihood.* IISD.

Ilham. (2013, January 21). *Penjaringan Masih Lumpuh.* http://poskotanews.com/ 2013/01/21/penjaringan-masih-lumpuh/. Accessed 4 Mar 2013.

IPCC – Intergovernmental Panel on Climate Change. (2000). *IPCC Special report on emissions scenario: Summary for policy makers.* Special Report on IPCC Working Group III. IPCC.

IPCC – Intergovernmental Panel on Climate Change. (2001). *Adaptation to climate change in the context of sustainable development and equity (Chapter 18),* IPCC Working Group III. Executive Summary. http://www.grida.no/publications/other/ipcc_tar/?src=/climate/ipcc_tar/wg2/642.htm

IPCC – Intergovernmental Panel on Climate Change. (2007). Climate change 2007: Impacts, adaptation and vulnerability. Contribution of Working Group II. In M. Parry, O. Canziani, J. Palutikof, P. van der Linden, & C. Hanson (Eds.), *The fourth assessment report of the Intergovernmental Panel on climate change.* Cambridge: Cambridge University Press.

IPCC – Intergovernmental Panel on Climate Change. (2012). Managing the risks of extreme events and disasters to advance climate change adaptation. In C. B. Field, V. Barros, T. F. Stocker, D. Qin, D. J. Dokken, K. L. Ebi, M. D. Mastrandrea, K. J. Mach, G.-K. Plattner, S. K. Allen, M. Tignor, & P. M. Midgley (Eds.), *Special report of Working Groups I and II of the intergovernmental panel on climate change.* Cambridge/New York.

ISSC/UNESCO. (2013). *World social science report 2013: Changing environments.* Paris: OECD Publishing and UNESCO Publishing.

IUCN, IISD, HELVETAS, & SEI. (2012). *CRiSTAL user's manual version 5: Community-based risk screening tool – Adaptation and livelihoods.* IUCN, IISD, HELVETAS, SEI.

Jakarta Post. (2009, January 8). BMG warns of big floods in January. *Jakarta Post.* 8 January 2009

Jamil, S., & Ali, M. (2013). Disasters in Southeast Asia's megacities: Protecting the informal sector. *Article in East Asia Forum.* http://www.eastasiaforum.org/2013/03/08/disasters-in-southeast-asias-megacities-protecting-the-informal-sector/. Accessed 8 Mar 2013.

Jellinek, L. (1999). Kampung culture or consumer culture. In *Consuming cities: The urban environment in the global economy after Rio.* London: Routledge.

JICA – Japan International Cooperation Agency-Research Institute. (2011). Climate change simulation study: The study on impact of climate-change-related flood on the urban poverty. *Household Survey Report, unpublished,* Jakarta.

Jordan, R. (2013). Jokowi Borong Lahan di Muara Baru untuk Relokasi Warga Waduk Pluit. *Detik.* http://news.detik.com/read/2013/05/09/161834/2241961/10/jokowi-borong-lahan-di-muara-baru-untuk-relokasi-warga-waduk-pluit

Julianery, B. E. (2007). *Kajian Upaya Pengendalian Banjir di DKI Jakarta: Rencana dan Realisasi Pembangunan Banjir Kanal Timur.* Tesis Pascasarjana Kajian pengembangan perkotaan. Jakarta: Universitas Indonesia.

Kantor Kelurahan Penjaringan. (2011). *Laporan tahunan 2011 Kelurahan Penjaringan Jakarta Utara.* Jakarta: Pemerintah Provinsi DKI Jakarta.

Kantor Kelurahan Penjaringan. (2012). *Laporan tahunan 2012 Kelurahan Penjaringan Jakarta Utara.* Jakarta: Pemerintah Provinsi DKI Jakarta.

Kantor Kelurahan Penjaringan. (2013). *Laporan tahunan 2013 Kelurahan Penjaringan Jakarta Utara.* Jakarta: Pemerintah Provinsi DKI Jakarta.

KBBI– Kamus Besar Bahasa Indonesia. (1991). *Bahasa, Pusat Pembinaan dan Pengembangan, Pusat Pembinaan dan Pengembangan Bahasa, Indonesia.* Jakarta: Departemen Pendidikan dan Kebudayaan, Balai Pustaka.

Kelly, P. M., & Adger, W. N. (2000). Theory and practice in assessing vulnerability to climate change and facilitating adaptation. *Climatic Change, 47*(4), 325–352.

Kemenkoeko – Kementerian Koordinator Perkonomian-RI. (2014, Oktober). *Pengembangan terpadu pesisir Ibukota negara: Masterplan Cetakan Pertama*. Jakarta

Kobayashi, K. (2007). *The invention of tradition in Java under the Japanese occupation: The Tonarigumi system and Gotong Royong*. Afrasian Centre for Peace and Development Studies Ryukoku University Working Paper Series No. 31.

Kompas. (2010). *Kampung di Jakarta*. Kompas 19 Februari 2010.

Kompas. (2011). Tanggul Raksasa disiapkan. *Kompas 7* December 2011. Available online at http://megapolitan.kompas.com/read/2011/12/07/05054187/Tanggul.Raksasa.Disiapkan

Kompas. (2013).Banjir dan Rob Rendam Jakarta. *Kompas* 14 January 2013. http://nasional.kompas.com/read/2013/01/14/02384166/banjir.dan.rob.rendam.jakarta

Korten, D. (1990). *Getting to the 21st century: Voluntary action and the global agenda*. Connecticut: Kumarian Press.

Korten, D., & Garner, G. (1984). Planning frameworks for people centered development. In D. C. Korten & R. Klauss (Eds.), *People center development: Contributions towards theory and planning framework* (pp. 201–209). West Hartford: Kumarian Press.

Kumar, S. (2002). *Methods for Community Participation: A complete guide for practitioners*. London: ITDG Publishing.

Kumar, C. (2005). Revisiting 'community' in community-based natural resource management. *Community Development Journal, 40*, 275–285.

Kundu, N. (2003). The Case of Kolkata, India. *Understanding Slums: Case Studies for the Global Report on Human Settlements 2003*. Development Planning Unit: UN-Habitat.

Kuruppu, N., & Liverman, D. (2011). Mental preparation for climate adaptation: The role of cognition and culture in enhancing adaptive capacity of water management in Kiribati. *Global Environmental Change, 21*(2), 657–669.

Kusno, A. (2000). *Behind the postcolonial: Architecture, urban space, and political cultures*. London/New York: Routledge.

Laukkonen, J., Blanco, P. K., Lenhart, J., Keiner, M., Cavric, B., & Njenga, C. (2009). Combining climate change adaptation and mitigation measures at the local level. *Habitat International, 33*(3), 287–292.

Laquian, A. A. (2004, June 24–25). *Who are the poor and how are they being served in Asian cities?*. Paper presented at the forum on urban infrastructure and public service delivery for the urban poor, regional focus, Asia, New Delhi, India.

Lian, K. K., & Bhullar, L. (2011). Governance on adaptation to climate change in the ASEAN region. *Carbon and Climate Change Law Review, 5*(1), 82–90.

Lobo, E. (2008). *Dampak Perubahan Iklim terhadap kondisi Banjir Jakarta. Tesis Pascasarjana Kajian pengembangan perkotaan*. Jakarta: Universitas Indonesia.

Lofland, J., & Lofland, L.H. (1999). Data logging in observation: Field notes. In A. Bryman and R.G. Burgess (Ed.), *Qualitative research*. London: Sage.

Lont, H. (2005). *Juggling money: Financial self-help organizations and social security in Yogyakarta*. Leiden: University of Passau.

Maguire, B., & Cartwright, S. (2008). *Assessing a community's capacity to manage change: A resilience approach to social assessment*. Canberra: Bureau of Rural Sciences.

Makmur, H. G. (2014, January 17). Musibah banjir menurut Islam. *Banjarmasih Post* Online. http://banjarmasin.tribunnews.com/2014/01/17/musibah-banjir-menurut-islam. Accessed 7 Feb 2014

Mäntysalo, R. (2004). *Approaches to participation in urban planning theories. Article of Department of Architecture*. Finland: University of Oulu.

Marcussen, L. (1990). *Third world housing in social and spatial development: The case of Jakarta*. Avebury: The University of Michigan.

Martin-Breen, P., & Anderies, J. M. (2011). *Resilience: A literature review*. The Rockefeller Foundation: New York.

McDowell, C., Canepa, C., & Ferreira, S. (2007). *Reflective practice: An approach for expanding your learning frontiers*. New York: MIT Open Course Ware.

McGranahan, G., Balk, D., & Anderson, B. (2007). The rising tide: Assessing the risks of climate change and human settlements in low elevation coastal zones. *Environment and Urbanization, 19*, 17–37.

McGray, H., Hammil, A., & Bradley, R. (2007). *Weathering the storm: Options for framing adaptation and development.* Washington, DC: World Resource Institute.

McLean, J. D., Freitas, C. R., & Carter, R. M. (2009). Influence of the southern oscillation on tropospheric temperature. *Journal of Geophysical Research: Atmosphere, 114(D14)*.

Meliana, T. (2005). Studi Daerah Rawan Genangan Di Jakarta Utara Akibat Kenaikan Paras Muka Laut Dan Penurunan Muka Tanah Di Teluk Jakarta. Tugas Akhir Strata-1. Program Studi Oseanografi, Institut Teknologi Bandung.

Mercy Corps. (2011). Local Resilience Action Plan (LRAP) Report from 3 Pilot Kelurahan: Technical assistance for climate resilient cities by implementing Kelurahan Empowerment Initiative (KEI) – DKI Jakarta. Jakarta: Mercy Corps Indonesia.

Mercy Corps Indonesia. (2010). *Community Profile of RW 17 Penjaringan.* Survey Report. Jakarta: Mercy Corps Indonesia.

Mielke, K. (2007). On the concept of Village in Northeastern Afghanistan: Explorations from Kunduz province. *ZEF Working Paper Series 70/Amu Darya Project Working Paper No. 6.* Bonn.

Mielke, K., & Hill, J. (2011). Interviewing: Teaching Material in the Disciplinary Program of ZEFa, Bonn, 13 January 2012.

Mishler, E.G. (2000). Validation in inquiry-guided research: The role of exemplars in narrative studies. In B.M. Brizuela, J.P. Stewart, R.G. Carrillo, and J.G. Berger (Ed.), *Acts of inquiry in qualitative research.* Cambridge, MA: Harvard Educational Review.

Moench, M., Tyler, S., & Lage, J. (2011). *Catalyzing urban climate resilience: Applying resilience concepts to planning practice in the ACCCRN program (2009–2011).* Boulder: Institute for Social and Environmental Transition, International.

Moser, C. O. N. (1978). Informal sector or petty commodity production: Dualism or dependence in urban development? *World Development, 6*(9/10), 1799–1825.

Moser, C. O. N. (1998). The asset vulnerability framework: Reassessing urban poverty reduction strategies. *World Development, 26*(1), 1–19.

Moser, C., & Satterthwaite, D. (2009). Towards pro-poor adaptation to climate change in the urban centres of low- and middle-income countries. Human settlements discussion paper series. London: IIED (International Institute for Environmental Development).

Muhammad, A. (2011). Kebijakan Nasional Perubahan Iklim; Sinergisitas Rencana Aksi Nasional – Daerah. Bahan presentasi pada kegiatan"Training of Trainers Mitra Bahari: Perencanaan Adaptasi Perubahan Iklim di Wilayah Pesisir dan Pulau-pulau Kecil, Ditjen Kelautan Pesisir dan Pulau-pulau Kecil (KP3K). Jakarta.

Narso, B. (2012). *Tanda dan Ciri Pranata wangsa warisan budaya Indonesia.* http://senijawakuno.blogspot.de/2012/12/tanda-dan-ciri-pranata-mangsa-warisan.html

Nelson, D. R., Adger, W. N., & Brown, K. (2007). Adaptation to environmental change: Contributions of a resilience framework. *Annual Review of Environment and Resources, 32*(1), 395.

Neuman, W. L. (2000). Social research methods: Qualitative and quantitative approaches. *Allyn and Bacon* (4th edition). Boston.

Neuman, W.L. (2006). Social research methods: *Qualitative and quantitative approaches.* Pearson.

Nicholls, R.J., Hanson, S., Herweijer, C., Patmore, N., Hallegatte, S., Morlot, J.C., Chateau, J., & Wood, R.M. (2007). Ranking of the world's cities most exposed to coastal flooding today and in the future. *OECD Environment Working Paper.*

O'Brien, K. (2013). *What's the problem? Putting global environmental change into perspective.* World Social Science Report 2013: Changing Global Environments ISSC & UNESCO.

Oberkircher, L., & Hornidge, A. K. (2011). Water is life'—Farmer rationales and water saving in Khorezm, Uzbekistan: A Lifeworld analysis. *Rural Sociology, 76*(3), 394–421.

Ooi, G. L., & Phua, K. H. (2007). Urbanization and slum formation. *Journal of Urban Health, 84*(1), 27–34.

Overseas Development Institute (ODI). (2013, October). *The geography of poverty, disasters and climate extremes in 2030*. ODI, UK. http://www.odi.org/sites/odi.org.uk/files/odi-assets/publications-opinion-files/8633.pdf

Pahl-Wostl, C. (2009). A conceptual framework for analysing adaptive capacity and multi-level learning processes in resource governance regimes. *Global Environmental Change, 19*(3), 354–365.

Parlindungan, C. (2003). *Tanah Telantar Menurut Peraturan Pemerintah Republik Indonesia No. 36 Tahun 1998 dan Permasalahan-permasalahan yang terdapat di lapangan, Bagian Hukum Administrasi Negara*. Fakultas Hukum: Universitas Sumatera Utara.

Patton, M.Q. (1980). *Qualitative evaluation methods*. Beverly Hills: Sage.

PEACE - Pelangi Energi Citra Enviro. (2007). Indonesia and Climate Change: Current Status and Policies. Full Technical Report. Jakarta: World Bank.

Pelling, M., High, C., Dearing, J., & Smith, D. (2008). Shadow spaces for social learning: A relational understanding of adaptive capacity to climate change within organisations. *Environment and Planning A, 40*(4), 867–884.

Perkasa, V., & Hendyotio, M. K. (2003). Inventing participation: The dynamics of PKK, Arisan and Kerja bakti in the context of urban Jakarta. In N. Yoshihara & R. D. Dwianto (Eds.), *Grass roots and the neighborhood associationsm*. Jakarta.

Pettengell, C. (2010, April). *Climate change adaptation: Enabling people living in poverty to adapt*. Summary Oxfam International Research Report. Oxfam GB.

Prameshwari, P. (2009). *The tides: Efforts never end to repel an invanding sea*. Jakarta Globe 24 July 2009. Available online at http://thejakartaglobe.beritasatu.com/archive/the-tides-efforts-never-end-to-repel-an-invading-sea/. Accessed 20 Mar 2013.

Prasad, N., Ranghieri, F., Shah, F., Trohanis, Z., Kessler, E., & Sinha, R. (2009). *Climate resilient cities: A primer on reducing vulnerabilities to disasters*. The International Bank for Reconstruction and Development, The World Bank.

Priyambodo, B. (2005). Dinamika Banjir Jakarta yang dipengaruhi Sea Water Level Rising dan Land Subsidence. Desertasi Doktor Sipil ITB.

PvW – Partners voor Water. (2010). *Jakarta resilient toward a coastal defence strategy*. Summary Project Description.

Quiano, K., & Mullen, J. (2013, January 18). *Indonesia floods kill 15; thousands flee raging waters in capital*. CNN http://www.cnn.com/2013/01/18/world/asia/indonesia-jakarta-floods/. Accessed 20 Jan 2013.

Rabe, P.E. (2011). Urban planning and climate change in Indonesia. *Final Report of the Roundtable Forum of Pacific Rim Council on Urban Development*. Palembang: PRUCD.

Rahmat, A. (2011). *Model Adaptasi Kota Pesisir Terhadap Pengaruh Kenaikan Permukaan Air Laut, Studi Kasus: Kota Jakarta Utara, Provinsi DKI Jakarta*. Master, Urban Studies, Universitas Indonesia.

Ratag, M.A. (2001). Model Iklim Global dan Area Terbatas serta Aplikasinya di Indonesia. Seminar Sehari Peningkatan Kesiapan Indonesia dalam Implementasi Kebijakan Perubahan Iklim. Bogor.

Redaksi Sains. (2011, December 12). Tidak ada banjir siklus lima tahunan. *Kompas Online*. http://sains.kompas.com/read/2011/12/12/20281896/BMKG.Tak.Ada.Siklus.Banjir.Lima.Tahunan. Accessed 8 Mar 2012.

Reid, R. D., & Sanders, N. R. (2010). *Layout planning Chapter 10. Operation management: An integrated approach*. Wiley.

Revianur, A. (2013, January 18). BNPB: Banjir Paling Parang di Kampung Melayu. Kompas Online http://sains.kompas.com/read/2013/01/18/13543180/BNPB.Banjir.Paling.Parah.di.Kampung.Melayu

Ritzer, G. (2011). *Sociological theory*. New York: McGraw-Hill.

Roy, A. (2005). Urban informality: Toward an epistemology of planning. *Journal of the American Planning Association, 71*(2), 147–158.

Roy, M., Jahan, F., & Hulme, D. (2012). Community and institutional responses to the challenges facing poor urban people in Khulna, Bangladesh in an era of climate change. *BWPI working paper 163*. Brooks World Poverty Institute, University of Manchester.

Rujak Center. (2011). Kata Fakta Jakarta. In E. Sutanudjaja, & A. Arifin (Eds.). Jakarta: Rujak Center for Urban Studies.

Rulistia, N. D. (2012, November 23). Changing slums into multistory kampung. *Jakarta Post*.

Sagala, S., & Damayanti, K. (2010). Social risk perception and community adaptation: Managing climate change impacts at coastal area of Jakarta. In *Joint World Conference on Social Work and social development 2010*. Hongkong.

Samsuhadi. (2010, October 2). Ground water utilization in the Jakarta basin. Paper presented to the discussion on concepts and strategy of sustainability future of Jakarta, Tarumanagara University, Jakarta, 2 October 2010.

Satterthwaite, D. (2007). *Adapting to climate change in urban areas: The possibilities and constraints in low-and middle-income nations* (Vol. 1). London:IIED.

Satterthwaite, D., Dodman, D., & Bicknell, J. (2009). Adapting cities to climate change: Understanding and addressing the development challenges (Earthscan climate). In D. Satterthwaite, D. Dodman & J. Bicknell (Ed.). Earthscan.

Schipper, L. F. (2007). Climate change adaptation and development: Exploring the linkages. *Tyndall Centre for Climate Change Research Working Paper, 107*, 13.

Schipper, L. F. (2010). *The Changwon declaration: Cities responding to climate change*. Changwon: UN Habitat-Government of Changwon.

Schon, D. A. (1982). Some of what a planner knows a case study of knowing-in-practice. *Journal of the American Planning Association, 48*(3), 351–364.

Schon, D.A. (1983). The reflective practitioner. *How Professionals Think In Action*.

Schön, D. (1987). *Educating the reflective practitioner: Toward a new design for teaching and learning in the professions*. San Fransisco: Jossey-Bass.

Schreurs, M. A. (2008). From the bottom up: Local and Subnatinal climate change politics. *The Joutnal of Environment and Development, 17*(4), 343–345.

Schütz, A. (1967). *The phenomenology of the social world*. Evanston: Northwestern University Press.

Schütz, A.L. (1973). *The structures of the life-world* (Vol. 1). Evanston: Northwestern University Press.

Schütz, A., & Luckmann, T. (1974). *The structures of the life-world*. London: Heinemann Educational Books.

Scott, W.R. (1992). *Organizations: Rational, Natural and Open Systems*: Pearson Education (US); International 2 Revised edition (Mar. 1992).

Scott, J. C. (1998). *Seeing like a state: How certain schemes to improve the human condition have failed*. Yale University Press.

Sentosa, A. (2010). *Social economic vulnerability as result of sea level rise in the Penjaringan sub-district, North Jakarta (case study: RW 01 Pluit)*. Master, Urban Studies, Universitas Indonesia.

Shah, F., & Ranghieri, F. (2012). *A workbook on planning for urban resilience in the face of disasters adapting experiences from Vietnam's cities to other cities*. Washington, DC: World Bank.

Sihombing, A. (2010). *Conflicting images of Kampung and Kota in Jakarta*. Saarbrücken: LAP Lambert Academic Publishing.

Silas, J. (1990). *Pembangunan Permukiman Bertumpu Pada Masyarakat*. Bandung: Gunadarma.

Silas, J. (1993). *Surabaya 1293–1993: A city of partnership*. The City Government of Surabaya.

Silver, C. (2008). *Planning the megacity – Jakarta in the twentieth century*. London: Routledge.

Simarmata, H. A., & Suryandaru, R. (2015). Institution and planning: A self-reflection from the disaster management planning in Indonesia. In B. Werlen (Ed.), *Global sustainability in cultural perspective: Challenges for trans-disciplinary integrated research*. Cham: Springer Press.

Smit, B., & Pilifosova, O. (2007). Adaptation to climate change in the context of sustainable development and equity. In *Chapter 18*. UNU-EHS Publication.

Smit, B., Burton, B., Klein, R. J. T., & Wandel, J. (2000). An anatomy of adaptation to climate change and variability. *Climatic Change, 45*, 223–251.

Smithers, J., & Smit, B. (1997). Human adaptation to climatic variability and change. *Global Environmental Change, 7*(2), 129–146.

Snidvongs, A. (2010). *Climate change vulnerability assessment and urban development planning for Asian Coastal Cities*, Final report submitted to APN-Project Reference Number: CIA2009-01-Snidvongs

Snow, D.A., & Benford, R.D. (1988). Ideology, frame resonance, and participant mobilization. *International Social Movement Research,1* (1), 197–217.

Sperling, F., & Szekely, F. (2005). *Disaster risk management in a changing climate*. Discussion paper prepared for the world conference on disaster reduction on behalf of the Vulnerability and Adaptation Resource Group (VARG). Reprint with Addendum on Conference outcomes. Washington, DC. Available online at http://unpan1.un.org/intradoc/groups/public/documents/apcity/unpan029300.pdf

Soehoed, A. R. (2004). *Membenahi Tata Air Jabodetabek: Seratus Tahun Dari Bandjir Kanaal hingga Ciliwung Floodway*. Jakarta: Djambatan.

Steffen, W., Crutzen, P.J., & McNeill, J.R. (2001). *Adaptation to climate change in the context of sustainable development and equity (chapter 18)*, IPCC Working Group III. Executive Summary. http://www.grida.no/publications/other/ipcc_tar/?src=/climate/ipcc_tar/wg2/642.htm

Steffen, W., Crutzen, P. J., & McNeill, J. R. (2007, December). The anthropocene: Are humans Now overwhelming the great forces of nature?. *JSTOR 36*(8). Royal Swedish Academic Science

Stermann, J. D. (2006). Learning from evidence in a complex world. *American Journal of Public Health, 96*(3), 505–514.

Stiftel, B. (2000). Planning theory. In R. Pelaseyed (Ed.), *The national AICP examination preparation course guidebook*. Washington, DC: AICP.

Strang, V. (2005). Common senses water, sensory experience and the generation of meaning. *Journal of Material Culture, 10*(1), 92–120.

Struyk, R. J., Hoffman, M. L., & Katsura, H. M. (1990). *The market for shelter in Indonesian cities*. Urban Institute Press.

Sturm, B. (2007). *Information in Life* [Digital Video Series]. University of North Carolina at Chapel Hill: School of Information and Library Science.

Sudaryono. (2012). *Fenomenologi Sebagai Epistemologi Baru dalam Perencanaan Kota dan Permukiman*. Pidato Pengukuhan Jabatan Guru Besar pada Fakultas Teknik Universitas Gajah Mada, Universitas Gajah Mada.

Suparlan, P. (2004). *Masyarakat & Kebudayaan Perkotaan; Perspektif Antropologi Perkotaan*. Jakarta: YPKIK.

Supriyanto, S. (2010). Kriteria tanah terlantar dalam peraturan perundangan di Indonesia. *Journal Dinamika Hukum, 10*(1).

Susandi, A. (2009, October 12–14*)*. *Integration of Adaptive Planning Across Economic Sector*. NWP technical workshop on integration of approaces to adaptation planning, Bangkok, Thailand.

Tapsell, S., McCarthy, S., Faulkner, H., & Alexander, M. (2010). *Social vulnerability and natural hazard*. London: Flood Hazard Research Center – FHRC Middlesex University.

Teitz, M. B. (2007). Planning and the new institutionalisms. In N. Verma (Ed.), *Institutions and planning: Analogical inquiry, current research in urban and regional studies*. Amsterdam: Elsevier.

Thakoerdin, O. S. J. (2009). *Climate change adaptation Handbook for mayors*. Background Paper, World Bank.

Toyudho,E.S. (2014). Kerugian Banjir di Jakarta Utara Rp 100 M per Hari. *Tempo*. http://www.tempo.co/read/news/2014/01/20/083546419/Kerugian-Banjir-di-Jakarta-Utara-Rp-100-M-per-Hari

Turner, N. J., & Clifton, H. (2009). It's so different today: Climate change and indigenous lifeway in British Columbia, Canada. *Global Environmental Change, 19*(2), 180–190.

Turner, B. L., Kasperson, R. E., Matson, P. A., McCarthy, J. J., Corell, R. W., Christensen, L., Eckley, N., Kasperson, J. X., et al. (2003). A framework for vulnerability analysis in sustainability science. *Proceedings of the National Academy of Sciences, 100*(14), 8074–8078.

Tyler, S., Reed, S.O., Macclune, K., & Chopde, S. (2010). *Framework and examples from the Asian cities climate change resilience network*. ISET.

UN – Habitat, United Nation Human Settlement Program. (2003). *The challenge of slums. Global report on human settlements 2003.* UN Habitat.

UN Habitat. (2010). *The changwon declaration: cities responding to climate change.* Korea: UN Habitat – Government of Changwon.

UN Habitat. (2013). *State of the world's cities 2012/2013: Prosperity of cities.* Kenya: UN-Habitat – United Nation Human Settlement Program.

UNDESA-United Nations Department for Economic and Social Affairs. (2003). *Guidelines for reducing flood losses: a contribution to the International Strategy for Disaster Reduction.* US NOAA & UN DESA publication. Available at http://www.unisdr.org/files/558_7639.pdf

UNDP – United Nation Development Programme. (2004). *Reducing disaster risk: A challenge for development.* New York: UNDP.

UNDP – United Nation Development Programme. (2008) Adapting to the inevitable: National action and international cooperation. *Fighting climate change: Human Solidarity in a Divided World,* UNDP. http://hdr.undp.org/en/reports/global/hdr2007-2008

UNDP. (2011). *Sustainability and equity: A better future for all.* UNDP.

UNDP. (2012). Annual report for PIMS: 3508-Community based adaptation to climate change (GEF SPA-Global FSP). UNDP. Available at http://www.adaptationundporg/sites/default/files/downloads/annual_report_of_2012_aj-cn_2_0.pdf

UNDP. (2013). *Community based resilience assessment (CoBRA) conceptual framework and methodology.* New York: Dry Lands Development Centre, UNDP.

UNEP. (1998). Handbook on methods for climate impact assessment and adaptation strategies. In J. Feenstra, I. Burton, J. Smith, & R. Tol (Ed.), *United Nations environment program, institute for environmental studies* (pp. 395). Amsterdam.

UNESCAP. (2007). *Statistical yearbook for Asia and the Pacific.* New York: UNESCAP.

UNFCCC (2009). *Adaptation assessment, planning and practice: an overview from the Nairobi work programme on impacts, vulnerability and adaptation to climate change.* In United Nations Framework Convention on Climate Change (Ed.). Bonn: UNFCCC.

UNFCCC – United Nations Framework Convention on Climate Change. (2010). *Fact sheet: The need for adaptation.* Bonn: UNFCCC.

UNISDR – United Nations International Strategy for Disaster Reduction. (2004). *Living with risk: A global review of disaster reduction initiative.* Geneva: UNISDR Report.

UNU-EHS. (2011). *Climate change and fragile states: Rethinking adaptation.* In United Nations University (Ed.), UNU-EHS.

USAID. (2009). *Adapting to coastal climate change: A guidebook for development planners.* USAID.

Van Dillen, S. (2004). *Different choices: Assessing vulnerability in a South Indian village.* Verlag für Entwicklungspolitik.

Verma, N. (2007). *Institutions and planning: Current research in urban and regional studies.* Netherlands: Elsevier.

Wagner, H. R. (1970). Alfred Schütz on phenomenology and social relations. In H. R. Wagner (Ed.), *The heritage of sociology.* Chicago: The University of Chicago Press.

Ward, P. J., Marfai, M. A., Yulianto, F., Hizbaron, D. R., & Aert, J. C. J. H. (2010). *Coastal inundation and damage exposure estimation: A case study for Jakarta.* Natural Hazard.

Watson, R.T., Zinyowera, M.C., & Moss, R.H. (1996). Technologies, policies and measures for mitigating climate change. *Technical Paper for IPCC working group 2.*

Weichhart, P. (2010). *Social geography and spatial planning: On people and life-world rationality in the planning process: Speech notes.* Salzburg. http://socgeo.ruhosting.nl/colloquium/Weichhart2.pdf

Wilhelm, M. (2011). The role of community resilience in adaptation to climate change: The urban poor in Jakarta, Indonesia. *Resilient Cities Journal,* 45–53. Springer.

Wilson, E. (2007). Comment: Response to interface 7(2): Is the issue of climate change too big for spatial planning? *Planning Theory & Practice,* 8(1), 125–127.

References

Winarno, H. (2013, January 17). BNPB: 5 Orang tewas akibat banjir besar di Jakarta. *Merdeka Online.* https://www.merdeka.com/peristiwa/bnpb-5-orang-tewas-akibat-banjir-besar-di-jakarta.htmlr. Accessed 7 Feb 2013.

Winayanti, L., & Lang, H. C. (2004). Provision of urban services in an informal settlement: A case study of *Kampung* Penas Tanggul, Jakarta. *Habitat International, 28*(1), 41–65.

Wiryomartono, A. B. P. (1995). *Seni Bangunan dan Seni Binakota di Indonesia: Kajian mengenai konsep, struktur, dan elemen fisik kota sejak peradaban Hindu-Buddha, Islam hingga sekarang.* Jakarta: PT. Gramedia.

Wisner, B., Blaikie, P., Cannon, T., & Davis, I. (2004). *At risk-natural hazards, people's vulnerability and disasters* (2nd ed.). London: Routledge.

World Bank. (2004). *Understanding the economics and financial impacts of natural disasters.* Disaster Risk Management Series No. 4. The World Bank, DC. Available at http://documents.worldbank.org/curated/en/146811468757215744/pdf/284060PAPER0Disaster0Risk0no-04.pdf

World Bank. (2010). *Jakarta: Urban challenges in a changing climate, a Mayors' task force on climate change, disaster risk, and the urban poor.* Jakarta: The World Bank Indonesia.

World Bank. (2011). *Climate change, disaster risk, and the urban poor: Cities building resilience for a changing world.* Washington: World Bank.

Yusuf, O. (2013). Ini dia, peta dijital resmi banjir Jakarta. *Kompas Online* 17 January 2013. Available online at http://tekno.kompas.com/read/2013/01/17/15432676/Ini.Dia.Peta.Digital.Resmi.Banjir.Jakarta. Accessed 6 Feb 2013.

Yusuf, A.A., & Fransisco, H. (2009). Climate change vulnerability mapping for Southeast Asia. *EEPSEA Economy and Environment Program for Southeast Asia.*

Zein, H.M. (2001). *Adaptation to climate change in the context of sustainable development and equity (Chapter 18).* IPCC Working Group III. Executive Summary. http://www.grida.no/publications/other/ipcc_tar/?src=/climate/ipcc_tar/wg2/642.htm

Zein, H. M. (2013). Musibah Banjir. *Republika Online* 18 January 2013. Available online at http://www.republika.co.id/berita/duniaislam/hikmah/13/01/18/mgt07g-musibah-banjir . Accessed 5 Feb 2013.

Zhu, J., & Simarmata, H. A. (2015). Formal land rights versus informal land rights: Governance for sustainable urbanization in the Jakarta metropolitan region, Indonesia. *Science Direct, 43*, 63–73.

MIX
Papier aus verantwortungsvollen Quellen
Paper from responsible sources
FSC® C105338

If you have any concerns about our products,
you can contact us on
ProductSafety@springernature.com

In case Publisher is established outside the EU,
the EU authorized representative is:
**Springer Nature Customer Service Center GmbH
Europaplatz 3, 69115 Heidelberg, Germany**

Printed by Libri Plureos GmbH
in Hamburg, Germany